Horatio Nelson Chute

Elementary Practical Physics

A Guide for the Physical Laboratory

Horatio Nelson Chute

Elementary Practical Physics
A Guide for the Physical Laboratory

ISBN/EAN: 9783337279691

Printed in Europe, USA, Canada, Australia, Japan

Cover: Foto ©berggeist007 / pixelio.de

More available books at **www.hansebooks.com**

ELEMENTARY
PRACTICAL PHYSICS

A GUIDE FOR THE PHYSICAL LABORATORY

BY

H. N. CHUTE, M.S.

Teacher of Physics in the Ann Arbor High School

BOSTON, U.S.A.,
D. C. HEATH & CO., PUBLISHERS.
1890.

PRESS OF
Rockwell and Churchill
BOSTON

PREFACE.

THIS book has been written with the object of promoting the teaching of Physics by what is known as the Laboratory Method. It embodies the Experimental Course which has been found suitable for students of the Eleventh and Twelfth Grades at the Ann Arbor High School, where the author has had several years' experience in conducting large laboratory classes.

It consists of a series of carefully selected exercises, both qualitative and quantitative in character, in which full directions are given regarding the preparation of the apparatus, and the manner of conducting the experiments, together with numerous suggestions about methods of observing, note-taking, and making inferences from data. The work is not designed to supplant the ordinary text-book, as definitions and statements of principles have been wholly omitted from its pages. It is recommended that the study of some suitable elementary treatise on Physics be carried on in the usual way, the teacher performing in the presence of the class either the experiments therein described or equivalent ones selected from this guide. Such instruction should, in the opinion of the author, precede the introduction of the student to the laboratory.

It is a mistaken supposition, unfortunately prevailing largely among science teachers, that the Laboratory Method requires a student to discover for himself the important laws and truths of the physical world. The race has been centuries in reaching its present knowledge of Nature's laws, and it is evidently unreasonable to expect the untrained mind of a boy or girl to discover and formulate in one year, or even in several years, the principles discussed in an elementary book on Physics. The office of the school physical laboratory is not one of original discovery; but, in the first place, to put the student in the best possible position to see what he looks at, in order that fact-knowledge may be added to word-knowledge in the most impressive way by having the head and hand work and learn together; and, in the second place, to train the faculties to exactness of observation, independence and carefulness in forming conclusions, and good judgment in weighing evidence.

It is not necessary, neither is it advisable, that the whole subject of Physics be studied in a didactic way before a student enters on his experimental work. All that is required is that the class instruction on any subject or topic shall precede the student's study of that subject experimentally, for it will be found that only in this way will the complete significance of the experiment and its results be fully perceived, and intelligent work secured to the exclusion of that which is lifeless and mechanical. No fear need be entertained that the student's interest will be lessened in the least by a knowledge of what he is likely to see, obtained either from seeing the experiment performed by the teacher or from reading about it in the class manual. The author's experience for a number of years has convinced him that the opposite effect is almost invariably produced; that is, a livelier and more intelligent interest is kindled thereby. The student always finds that the close contact with the experiment in the laboratory reveals to him enough that he did not see when he witnessed it in the class-room, or found no mention of in the manual he consulted, to compensate him many times over for the time he now has put upon it.

There is, however, a large class of experiments that cannot be performed to any advantage outside of the laboratory, particularly those of a quantitative character, involving the verification of physical laws and the determination of physical constants. On these subjects but little help is furnished by the ordinary text-books, and students graduate every year from our High Schools and Colleges absolutely ignorant of the difficulties in the way of accurately determining the simplest physical constant, or establishing the truth of the most familiar physical law. To experiments of this kind special attention has been given in these pages, for it is believed that it is only by doing this so-called " dead-work " can a student ever come to a correct and complete understanding of these laws. The objection that such work is beyond the students who ordinarily study physics in our High Schools is fully met by the fact that the data appended to Exercises 275 and 276 of this book were taken from the note-books of two young girls who had no other aid than that furnished in the problems as given there.

Although in the directions accompanying the exercises considerable attention is given to the construction of apparatus, the author would not be understood as approving of converting the physical laboratory into a mechanical work-shop. The object in giving so many details was threefold. First, to make it possible for schools, having but little means for purchasing apparatus, to get together, in time, through the aid of teacher, students, and local mechanics, a supply of apparatus that has the one great merit, "it seldom fails to do that which is required of it." Secondly, to show that the verification and the illustration of Nature's great laws require no elaborately finished and expensively constructed appliances; but that, on the contrary, the more simple and inexpensive the mechanism is, the more satisfactory is usually its performance. Thirdly, to give the student as clear an idea as possible of the instrument placed in his hands by putting before him the specifications followed in its construction. The author well remembers that no study of books ever gave him as clear an idea of the induction coil as he got from a working-drawing and specifications of one published several years ago in a well-known scientific periodical.

Special attention is called to the appendices. In the first, the author has attempted to outline a method of conducting laboratory work: the equipment of a room, the selection of apparatus, and the management of large classes with limited facilities are some of the topics discussed at considerable length. In the second appendix there is described a large number of operations, such as teacher and student are often called on to perform, in repairing, constructing, and adjusting apparatus. Many very valuable recipes find a place there, all of which have been thoroughly tried, and are known to be most excellent. In the third appendix is given a very complete list of such Constants as will be needed or found convenient in the application of the truths of Physics to the solution of practical problems. They will also serve as guides to the student, often indicating to him what degree of accuracy he has secured in his experimentation.

It is too much to expect that these pages will be found free from errors; still it is believed that a fair degree of accuracy has been secured. Dr. T. C. Mendenhall, President of the Rose Polytechnic Institute, has carefully and critically read the whole work in manuscript and in proof, and has contributed many valuable suggestions. The following well-known science teachers have also carefully read the work in proof, and aided very greatly in giving it its present degree of accuracy and completeness of treatment: W. Le Conte Stevens, Packer Collegiate Institute, Brooklyn, N. Y.; J. Montgomery, Kalamazoo College, Kalamazoo, Mich.; E. P. Jackson, Latin School, Boston, Mass.; J. Y. Bergen, Latin School, Boston, Mass.; James H. Shepard, Agricultural College, Brookings, Dak.; George B. Merriman, Rutgers College, New Brunswick, N. J.; J. H. Pillsbury, Smith College, Northampton, Mass.; George N. Cross, Exeter Academy, Exeter, N. H.; George I. Hopkins, High School, Manchester, N. H.; A. C. Boyden, Normal School, Bridgewater, Mass.; George L. Chandler, High School, Newton, Mass.; Albert C. Hale, Boys' High School, Brooklyn, N. Y.; Arthur M. Mowry, High School, Salem, Mass.; and many others.

Special acknowledgments are due to A. P. Gage, of Boston, James W. Queen and Co., Philadelphia, and E. S. Ritchie and Sons, of Brookline, Mass., for the use of a number of illustrations. The author would also acknowledge his indebtedness to that great company of scientific workers who have furnished him, through their writings, with the material, and sometimes with the words, for the pages which follow. Should this imperfect effort prove itself helpful to that large body of earnest men and women who are striving to solve the great problem of "what is the best method of science teaching," the author will feel fully repaid for the many laborious hours of day and night, in laboratory and in study, he has spent for a number of years in collecting, verifying, selecting, and adjusting the matter contained in the laboratory guide which he now offers to the teaching profession.

TO THE TEACHER.

I.

HOW TO USE THIS MANUAL.

1. The plan on which this book is prepared does not restrict it to any particular theory of practical science teaching. If you believe that a full course of didactic instruction should precede the admission of students into the laboratory, and that the practical work should be entirely quantitative, then omit the qualitative and select from the quantitative such as is adapted to your wants. On this plan, however, it will generally be found very difficult to handle large classes, as it necessarily crowds all the practical work into a small part of the school year, rendering it nearly impossible to arrange a programme which provides enough hours per week for each student to enable him to accomplish much. It would be better to organize the laboratory work a week or so after the class-room work has begun, and carry on the two simultaneously. The didactic instruction will by this plan be in advance of the practical work sufficiently to enable the student to carry on his work intelligently. About three hours per week devoted to class recitations on some good text-book, accom-

panied by illustrative experiments by the teacher; two to six hours per week spent in practical work in the laboratory; and one hour per week spent by students in class, discussing and comparing their reports on this work, is the plan on which the author has taught physics for several years with results that are quite satisfactory.

2. If, on the other hand, you believe that the experimental work should precede that of the class-room, a view held by many of our ablest teachers, you will find that the Guide contains a very complete collection of experiments, from which you can select those adapted to your facilities. These experiments are accompanied by hints on observing, recording observations, and correctly interpreting data obtained, so that the labor of supervision will be greatly reduced, and the burden of preparing the work reduced to a minimum.

3. It will not interfere with the plan of the book to omit any exercise not adapted to the laboratory facilities. It is best to assign work to the student that he has the genius requisite to do something with. See Art. 596. The large variety of material from which to choose will make it possible to make changes each year. Old note-books will be of less value to new classes if this is done.

4. It may, occasionally, be found desirable to make changes in a problem to adapt it to a piece of apparatus of different design from that contemplated in the Guide. These changes may be explained in the work-room or class-room, as found most convenient.

5. Require students to use reference books freely in studying their practical work. Also encourage them to make changes in the mode of procedure. See Art. 599, Rule 4. It will not unfrequently happen that a student will hit on some little modification that will enable him to secure much more accurate results, mainly for the reason that he devised it, and therefore could handle it all the better on that account.

6. It will be seen that the book seldom states what the results should be. Experience has convinced the author that more independent and honest work can be secured on this plan. Let the average student take the ordinary text-book into the laboratory and work through one of its experiments, and in the majority of cases he will believe that he got exactly what the book account said he would, nothing more and nothing less, notwithstanding the fact that half of our book-makers, according to Prof. Tyndall, describe experiments which they never made, their descriptions often lacking both force and truth.

II.

HOW TO SECURE APPARATUS.

1. It is a mistake to suppose that practical work with large classes necessitates a large and expensive collection of apparatus. The cost of equipping a Physical Laboratory is less than is generally supposed. Nearly every school

already possesses the pieces that are expensive, as air-pump, electrical machine, and their many accessories. When the author decided to introduce the Laboratory Method in the Ann Arbor High School, he asked the School Board to do three things: furnish a room with gas, water, and a few flat-top tables; appropriate $100 to purchase a few of the simpler devices for making accurate measurements; and require each student to pay each term a small fee, the proceeds to be devoted to enlarging the equipment. The work began with over eighty students, and now for several years not a dollar has been asked of the School Board for purchasing apparatus; still the facilities are rapidly improving each year, and are probably superior to those of most schools of the kind.

2. In Art. 595 are given many hints regarding apparatus.

3. The majority of teachers have but little time to give to the construction of apparatus; still, in most cases, something can be done, especially in putting things together that were prepared by some mechanic. There is such work as winding the wire on galvanometer frames and making the needles for the same, winding electro-magnets and magnetizing steel magnets made by a local blacksmith, etc. School Boards are not easily induced to appropriate money to be expended outside of the district, but they will generally be found willing to employ a carpenter for a few days to work for the school. There are a great many devices described in these pages that such a man

can make, and just as satisfactorily as if he lived in some distant city. These instruments may require adjustment, a statement true of 90 per cent. of all apparatus purchased of the most noted makers.

4. Europe is the great storehouse of cheap apparatus, and under the law, schools can import for their own use duty free. In this way a saving of 30 to 50 per cent. can often be effected. This is especially true in the matter of lenses, Grenet batteries, balances, weights, etc. This importing can be done through any of the firms referred to in Art. 596.

III.

HOW TO MANAGE THE LABORATORY.

1. There are in use two methods of conducting laboratory work, known as the *separate system* and the *collective system*. Under the former each section of two students would work on different problems, the apparatus going around in rotation. It is difficult, under this plan, to have the student's work conform to a strictly logical order, but it requires little or no duplication of apparatus. The collective system is the ideal one. Under it all are engaged on the same work at the same time. It has this advantage over the separate system, a teacher can instruct all at once on any point demanding more than ordinary care, and will have more time to devote to the few who may be less apt in their work. The author would recommend a combining

of the two methods as better adapted to the circumstances of most schools. This would involve a duplication of those appliances which are less expensive, as thermometers, lenses, mirrors, galvanometers, etc.

2. In Arts. 596, 597, and 599 will be found quite full directions and suggestions on conducting practical work. Those which relate to order are very important. The student should understand that the work-room is not a play-room, and that every moment of his time must be put to good use.

3. By arranging the school programme so that students taking physics have all their recitations in the forenoon, it will be possible for them to devote at least two of their afternoons to the laboratory each week. If possible two consecutive hours should be spent there, for it will not unfrequently happen that one hour will be too short to complete a line of investigation, and to stop when not through is to lose all that has been done.

4. It is recommended that, generally, two students be permitted to work together on the same problem, as more than two hands are often required in performing an experiment. This plan will increase the number of students who can be kept at work. It is doubtful if one person can look properly after more than twelve students, and when large numbers necessitate admitting many more than this number to the room an assistant should be employed.

5. If the students have definite places assigned them in the room it will conduce to good order. Each student

should be held responsible for the apparatus put at his disposal, and if broken or in any way injured through his fault he should be required to make it good.

6. The work expected each week from the student should be assigned him a few days in advance, and he should be required to give it careful study before reporting for duty, in order that he may not be hampered by ignorance of the successive steps in the experiment. The author does not believe it is wise to restrict all experiments to those of a quantitative character. He would require the student to repeat for himself many of the experiments of the class-room, with the object of bringing him into actual contact with the things themselves, so that their properties and relations may become familiar as solid, first-hand mental acquisitions. Furthermore, there are many important truths, the establishment of which involves no exact measurements, that cannot be developed experimentally in the class-room without expensive apparatus, whereas a very simple device placed in the student's hands will enable him to see the whole matter clearly.

IV.

HOW TO SHORTEN THE COURSE.

1. As is implied in more than one place, this book contains much more work than a student can master in the time usually devoted to the subject. There is no reason

why a teacher should not omit any exercise or set of exercises he chooses. It was thought advisable to have the book fairly complete in its treatment of each subject, that it might the better meet the wants of the greater number, as it often happens that what is adapted to the needs of one school is not to those of another. All subjects, however, are not equally important, and circumstances will often compel the omission of certain questions. The author would suggest that generally the following topics might be omitted without seriously detracting from the value of the course: Secs. III., IV., V., VI., VIII., X., and XI. of Chap. I.; VII. of Chap. III.; IX. of Chap. IV.; XV., XVI., and XVII. of Chap. V.; VII., XVI., XVII., and XVIII. of Chap. VI.; X., XI., and XIII. of Chap. VII. From the other sections select such experiments as bring out most satisfactorily with the apparatus at command the principles they are designed to teach. Under many exercises are found several methods of accomplishing the same thing. It adds interest to the work to assign different methods to different students and have them compare results.

2. If both time and appliances are limited, a still further contraction is possible, and in that case the author recommends that some such selection as suggested in Art. 596 be made. This principle should govern the work at all times; a few experiments thoroughly studied will be much more instructive than a great many superficially and hastily gone over.

CONTENTS.

CHAPTER I.

	PAGE
THE PROPERTIES OF MATTER EXPERIMENTALLY DETERMINED .	1–68
I. Extension. — Measurements of Length, Area, and Volume	1–22
II. Estimation of Mass	23–32
III. Impenetrability	33–36
IV. Divisibility	36–38
V. Porosity	38–41
VI. Indestructibility	41–44
VII. Cohesion	44–50
VIII. Elasticity	50–54
IX. Capillary Action	55–60
X. Solubility	61–62
XI. Diffusion	63–68

CHAPTER II.

MECHANICS OF SOLIDS	69–100
I. Laws of Motion	69–77
II. Centre of Mass. — Stability	77–80
III. Curvilinear Motion	80–82
IV. Accelerated Motion. — Gravitation. — Projectiles,	83–92
V. The Pendulum	92–96
VI. Friction	96–97
VII. The Simple Machines	97–100

CONTENTS.

CHAPTER III.

		PAGE
MECHANICS OF FLUIDS		101–126
I.	Pressure in Fluids	101–108
II.	Law of Boyle	108–110
III.	Law of Pascal	110–113
IV.	The Siphon and Pump	113–116
V.	The Principle of Archimedes	116–119
VI.	Determination of Density	120–126

CHAPTER IV.

HEAT		127–169
I.	Heat, and Mechanical Motion	127–132
II.	Heat and Chemical Action	132
III.	Conduction of Heat	133–136
IV.	Convection of Heat	136–139
V.	Expansion by Heat	140–145
VI.	Thermometry	146–152
VII.	Radiant Heat	153–159
VIII.	Calorimetry	160–167
IX.	Artificial Cold	167–169

CHAPTER V.

MAGNETISM AND ELECTRICITY		170–237
I.	Magnets. — Polarity. — Induction	170–173
II.	Nature of Magnetism	173–174
III.	The Magnetic Field	174–179
IV.	Terrestrial Magnetism	179–180
V.	Frictional Electricity	180–184
VI.	Statical Induction	184–186
VII.	Electrical Distribution	187–190
VIII.	Condensers	190–193
IX.	Electrical Machines	193–196
X.	Voltaic Electricity. — The Battery	196–199
XI.	Effects of Electrical Currents	200–204
XII.	Electrical Measurements	204–221

CONTENTS.   XIX

MAGNETISM AND ELECTRICITY. — *Continued.*   PAGE
   XIII. Electro-Magnetism and Electro-Dynamics . . 221–227
   XIV.   Current Induction   227–232
   XV.   Luminous Effects   233–236
   XVI.   Thermo-Electricity   236–237

CHAPTER VI.

SOUND   238–282
   I.   Wave Motion   238–241
   II.   Sources of Sound   241–244
   III.   Transmission of Sound   244–245
   IV.   Velocity of Sound   245–248
   V.   Propagation of Sound   249–250
   VI.   Reflection of Sound   250–252
   VII.   Refraction of Sound   252
   VIII. Loudness of Sound   253–256
   IX.   Interference of Sound   256–258
   X.   Sympathetic Vibrations   258–260
   XI.   Pitch of Sound   261–262
   XII.   Laws of Vibrating Rods and Strings . . . 263–266
   XIII. Overtones   266–267
   XIV.   Laws of Sounding Air-Columns 268–270
   XV.   Harmony and Discord 270–271
   XVI.   Vibrating Plates and Bells 271–273
   XVII. Attraction of Vibrating Bodies   273
   XVIII. Graphic and Optical Study of Sound . . . 273–281
   XIX.   Vocal Organs   282

CHAPTER VII.

LIGHT 283–333
   I.   Sources of Light 283–286
   II.   Rectilinear Propagation of Light   . . . 287–289
   III.   Photometry 289–291
   IV.   Reflection of Light 291–297
   V.   Mirrors 297–305
   VI.   Single Refraction of Light 305–309
   VII.   Lenses 309–314

CONTENTS.

Light. — Continued.

		PAGE
VIII.	Dispersion	314–317
IX.	Color	317–320
X.	Spectrum Analysis	320–323
XI.	Interference of Light	324–325
XII.	Optical Instruments	325–328
XIII.	Double Refraction and Polarization of Light	329–333

APPENDICES.

A.	— The Physical Laboratory	337–353
B.	— Laboratory Operations	354–360
C.	— Tables for Reference	361–376
	I. Capillarity	361
	II. Densities of Various Substances	361
	III. Limit of Elasticity	364
	IV. Electrical Conductivity	364
	V. Approximate Electro-motive Force of Primary Batteries	365
	VI. Electrical Resistance, Diameter, etc., of Pure Copper Wire	365
	VII. Acceleration due to Gravity	367
	VIII. Heat, Absorbing, Conducting, Radiating, Reflecting Power	367
	IX. Boiling-points of Substances at Bar. Pres. 76 cm.	368
	X. Coefficients of Expansion	368
	XI. Latent Heat of Liquefaction and Vaporization	369
	XII. Melting Points	369
	XIII. Specific Heat	370
	XIV. Indices of Refraction	370
	XV. Mensuration Rules	371
	XVI. Length of Seconds' Pendulum	371
	XVII. Velocity of Sound at 0° C.	372
	XVIII. Elasticity of Traction	372
	XIX. Tenacity	373
	XX. Trigonometrical Functions	373
	XXI. Some Useful Numbers	375
	XXII. Weights and Measures	375

PRACTICAL PHYSICS.

CHAPTER I.

THE PROPERTIES OF MATTER EXPERIMENTALLY DETERMINED.

I. EXTENSION. — MEASUREMENTS OF LENGTH, AREA, AND VOLUME.

1. Apparatus. — The fundamental principle involved in all measurements is that of direct comparison with some assumed standard, the accuracy of the results depending upon that of the standard, and the methods adopted to insure close comparisons. Among the many instruments employed in this kind of investigation may be mentioned the **Dividers, Diagonal Scale, Metre-Rod, Verniered Steel Caliper, Micrometer Caliper, Spherometer, Outside and Inside Caliper, Graduated Measure, Beam Compass, Proportional Dividers, Protractor,** etc. These are described in the following articles of this section, their methods of use explained, and in many cases simple and efficient substitutes suggested.

2. Exercise. — Measure the distance between two points situated on any plane surface, as a sheet of brass or cardboard.

2 *PRACTICAL PHYSICS.*

This can be done with sufficient accuracy by means of a pair of **Dividers** (Fig. 1) and a **Diagonal Scale** (Fig. 3). **Spring Dividers** (Fig. 2) are preferable. Any accurately and finely divided scale can be used; but of cheap scales, the diagonal scale, boxwood or ivory, is the most suitable for this problem.

Open the dividers, and place one of its points *exactly* on one of the given points on the plane surface, and the other point on the second given point. Now place one point of the dividers at one of the divisions 1, 2, 3, etc., of the scale, as 2, being careful to select one that will cause the other point to fall at 0, or between 0 and the vertical line bounding the scale. If the right-hand point falls at 0, then the required distance is 2 units. If the right-hand point falls to the right of 0, as at the diagonal line 3, then the distance is 2.3 units. If the right-hand point falls between two of these diagonal lines, as 3 and 4, then move the dividers toward the opposite edge of the scale, keeping the line of the points parallel to the lines running lengthwise of the scale,
Fig. 1. and at the same time the left-hand point in   Fig. 2. the line 2, until the right-hand point meets the intersection of a diagonal line with a horizontal one. Should such a point be where the diagonal line 3 intersects the horizontal one 7, then the distance is 2.37 units. This multiplied by the value of the scale unit, expressed either in inches or centimetres, will give the absolute length.

In conveying the measurements to the scale, there is danger that the dividers may get closed up a little from pressure of the fingers in holding the instrument. This must be carefully guarded against. Again, the points of the dividers must be placed with accuracy on the given points, and no pressure applied that would be likely to indent the surface in the least. Several independent measurements must always be made; then, by adopting their arithmetical mean, a result differing less from the exact one is more likely to be obtained. These

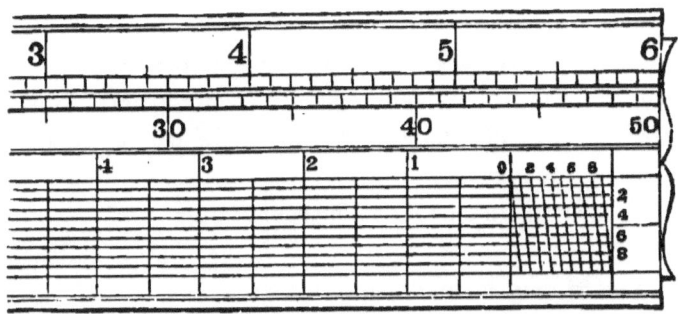

Fig. 3.

different measurements should all be neatly and accurately recorded at the time they are made, in a suitable note-book provided for the purpose, observing some such plan as the following : —

LINEAR MEASUREMENTS.

Problem. — Determination of distance between two points on cardboard. Sept. 5, 1888.

Method. — Dividers and scale.

Results. —

    First measurement . . . .
    Second   "   . . . .
    Third   "   . . . .

        Mean . . . .
        Mean in cm. . . . .
        Value of scale unit . . . .

3. Exercise. — Draw on paper three straight lines whose lengths are respectively 2.34 inches, 3.57 inches, and 1.89 inches.

4. Exercise. — Cut a circle, as accurately as possible, from cardboard. Measure its circumference by rolling it along a straight line drawn on paper. Measure its diameter, and compute the ratio of the circumference to the diameter.

A pair of dividers and a diagonal scale will be found convenient for the purpose.

Make several independent measurements and record the results.

The record may be made as follows: —

Problem. — Determination of the ratio of the circumference of a circle to its diameter. Sept. 5, 1888.

Method. — Dividers and diagonal scale.

Results. —

	Circumference.	Diameter.
First measurement . . .		....
Second " . . .		
Third " . . .	....	....
Mean	....	

Ratio, ... Error, ... Per cent of error, ...

The *error* is the difference between the ratio obtained and the true value, 3.1416.

5. Exercise. — Measure the length and the breadth of one of the tables in the room, employing both the English and the French measure. Reduce each to the other, then compare and point out in how many ways the differences in the final results may be accounted for.

THE PROPERTIES OF MATTER. 5

The metre-rod, such as is furnished by the American Metric Bureau, is a suitable instrument for solving this problem. The results may be recorded as follows: —

Problem. — Determination of the length and breadth of laboratory table No. . Sept. 5, 1888.
Method. — The metre-rod.
Results. —

	Length.		Breadth.	
First measurement . .	........cm.	........in.	........cm.	........in.
Second " . .	....			
Third " . .				
Mean	....	....	....	....
Reduced			....	....
Difference	....		....	....

6. Exercise. — Measure the length of a short rod or bar of wood or metal.

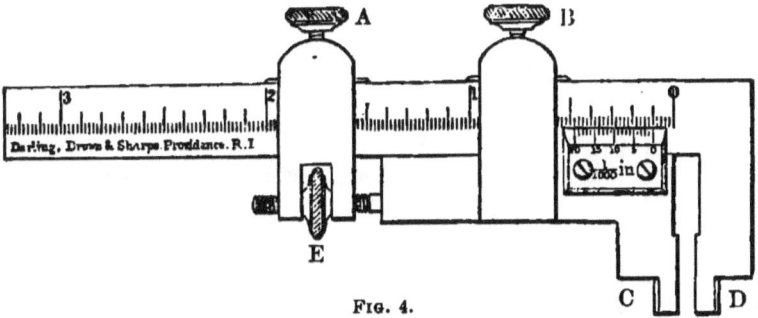

FIG. 4.

If great accuracy is required, it will be necessary to employ some such instrument as the **Verniered Steel Caliper**, an appliance so constructed that measurements as small as fiftieths of a millimetre can be made by its aid. Figs. 4, 5, and 6 exhibit some of the forms given it; the latter being without the **Vernier**, a device for estimating fractional parts of the smallest division on the scale. Where the verniered form cannot be had, for financial reasons, this simpler form may be used instead, generally with sufficient accuracy, if the fractional parts of the divisions are carefully estimated by the eye.

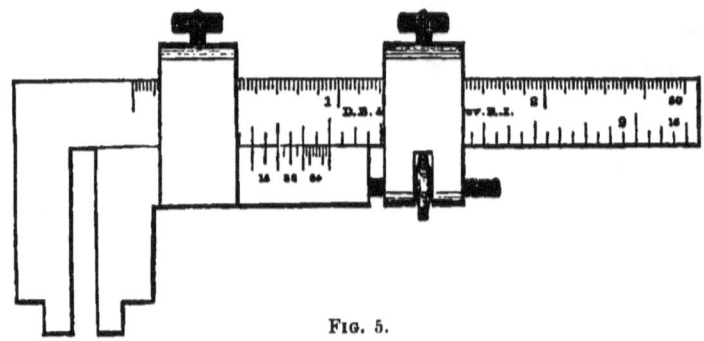

FIG. 5.

To measure the rod, unclamp the screws A and B (Fig. 4), and slide the jaw C away from D, so that the rod can be placed between C and D, with its axis *parallel* to the bar of the caliper. Then clamp A, and move the jaw C by the slow-

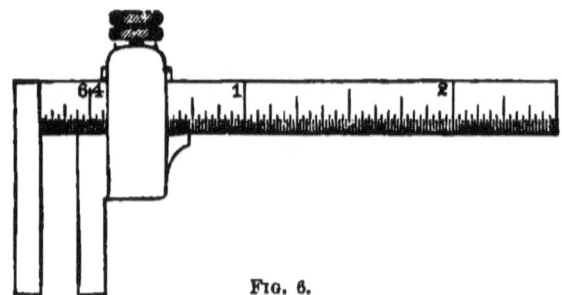

FIG. 6.

motion screw E till the rod is just firmly held between the jaws C and D, as it is an easy matter by springing the frame of the instrument, or compressing the rod, to change the reading one or more of the smallest units given by it. It now remains to read on the scale the length of the rod. On C will be seen a

THE PROPERTIES OF MATTER.

short scale, called a **Vernier**, 20 of whose divisions equal 19 on the bar or limb, the bar divisions being fiftieths of an inch. Hence, each division of the vernier is $\frac{1}{1000}$ of an inch shorter than each bar division. Therefore, write down the number of inches and fiftieths of an inch up to the zero of the vernier scale; then if the zero of the vernier does not register exactly with a division on the bar, observe which one of the vernier divisions coincides with a division on the bar, and this gives the number of thousandths to be added to the quantities already obtained. Express all fractional parts decimally. Make several independent measurements, and obtain their mean. Record the results after the manner described in previous exercises.

The index on the opposite side of this instrument is arranged for both inside and outside measurements. The outer edges of the jaws are rounded for insertion in tubes, or hollow cylinders; then, by reading from the index marked *inside*, the distance between the outer surfaces of the jaws is given, that is, the diameter of the tube.

This caliper can also be used for measuring the diameter of a cylinder, the diameter of a small ball, or the thickness of plates.

7. Exercise. — Determine the dimensions of a piece of brass tubing. Compute the lateral surface and the volume.

The best instrument for this purpose is the **Verniered Caliper**. Fair results can be got by the direct application of a good scale.

Enter results in the note-book as follows: —

Problem. — Determination of the dimensions of a piece of brass tubing. Sept. 7, 1888.

Method. — Verniered caliper.

Results.—

	Length.	Outside Diameter.	Inside Diameter.
First measurement. .	........in.	........in.	........in.
Second " . .	....		
Third " . .	....		....
Mean		....	....
Computed area = ...sq. in.		Volume = ...cu. in.	

8. Exercise. — Cut from a piece of wire the following lengths: 5 mm., 1.26 inches, 2.39 inches, 149 mm., 3.06 inches.

With a pair of **Cutting Pliers** (Fig. 7), or a file, cut off a piece of wire a little longer than is required. Set the

FIG. 7.

verniered caliper to the required length, and then, with a fine file, shorten the wire till it fits accurately between the jaws. A flat scale may be used in place of the caliper by laying the wire directly on the scale, carefully avoiding parallax in reading by always sighting across the ends of the wire perpendicularly to the scale.

9. Exercise. — Measure the diameter of a wire, and also determine the gauge-number.

The diameter can be found by means of the verniered caliper (Fig. 4), but a better instrument for the purpose is the **Micrometer Caliper** (Fig. 8). In using this appliance, it is necessary to find the pitch of the screw C by observing the graduation of the linear scale a. This is usually either

THE PROPERTIES OF MATTER. 9

fiftieths of an inch, or half-millimetres. The circular scale on D is generally divided into 20 parts when the graduation is in English measure, and 25 when French; so that a circular division represents $\frac{1}{20} \times \frac{1}{50} = \frac{1}{1000}$ of an inch in the one case, and $\frac{1}{2} \times \frac{1}{25} = \frac{1}{50}$ of a millimetre in the other. As these circular divisions are large, it is easy to estimate to quarters of these spaces. When the screw C is in contact with B, the zero of the scale on D should be at the zero of the linear scale a. If this is not the case, the set-screw which holds B should be changed till the zeroes of the two scales coincide.

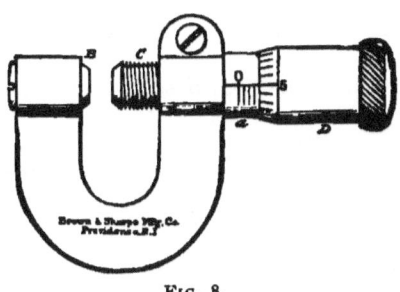

FIG. 8.

Now place the wire to be measured between B and C, and, by turning D, advance C till the wire is held between B and C; being careful not to exert undue pressure, as it is an easy matter to vary the reading one or more divisions of the circular scale by springing the frame of the instrument. To get the reading, observe how many divisions on the linear scale a are exposed, and what division of the circular scale on D coincides with the line running lengthwise of the cylinder having the linear scale. As shown in Fig. 8, we have the following reading: —

        On the linear scale, 4 small divisions = .08 inch.
        On the circular scale, 5 divisions   = .005 inch.
            Total reading   = .085 inch.

When the linear scale is graduated to read to fortieths of an inch, the larger divisions being tenths, then the circular scale is divided into 25 parts, and hence the reading is to thousandths.

A simple method of measuring the diameter of a wire is to wrap it *closely* around a cylinder whose diameter is large compared with that of the wire, till some 25 or more convolutions are obtained. Then the diameter will equal the width of the surface covered divided by the number of turns.

The diameter can also be determined by weighing accurately a known length of the wire, first having made the ends square with a file. Then the diameter in centimetres $= \sqrt{\dfrac{w}{.7854 dl}}$, in which $w =$ weight in grammes, $d =$ density (see Table II.), and $l =$ length in centimetres. For an explanation of this formula, consult the subject of density as presented in any treatise on Physics.

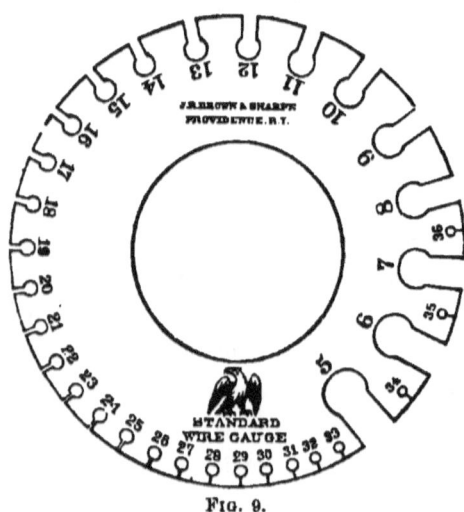

Fig. 9.

To ascertain the gauge-number, a **Wire-Gauge** (Fig. 9) is employed. This consists of a steel plate with a graduated series of notches around the edge, each one numbered according to some specified table of wire-gauges. Find, by trial, into which notch the wire will just fit, and read off the number on the gauge. By reference to Table VI., the diameter can be approximately found.

10. Exercise. — Measure the thickness of specimens of the following: Paper, mica, tin-foil, tinsel, etc.

The most suitable instrument for this problem is the micrometer caliper. Make several measurements of each specimen, and devise some convenient form for recording the results.

11. Exercise. — Measure the thickness of a very thin piece of mica.

This can be done by the aid of the micrometer caliper; but a more desirable instrument for making such measurements is the **Spherometer** (Fig. 10), as it is less likely to break the thin plate, and the moment of contact of the parts of the instrument with the substance can be more accurately determined, as the phenomenon is independent of the delicacy of the sensation of touch. This instrument consists of a metal platform supported by three steel legs whose extremities are the vertices of an equilateral triangle.

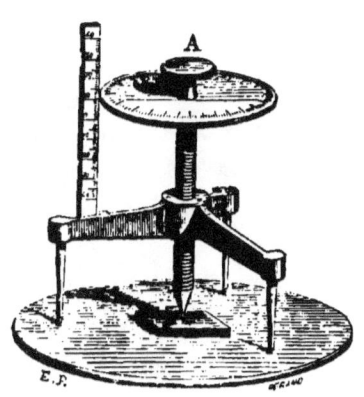

Fig. 10.

In the middle of this triangle is a fourth leg, which can be raised or lowered by means of a fine thread cut upon it. This leg passes through the platform, and terminates above in a finely graduated disk of such a size as to be tangent to a straight scale which is normal to the plane of the triangle. A true glass plane must accompany the instrument. In the absence of any thing better, a piece of double-thick French plate glass may be used. When the spherometer is placed on the plane with the four legs touching it, the zero on the disk should be exactly at the zero of the straight scale.

If found not to be in adjustment, loosen the screw A, and move the disk the necessary amount. In reading the instrument, the pitch of the screw must be known, and also the number of divisions on the disk. Let us suppose that the pitch is $\frac{1}{2}$ mm., and that the disk is divided into 500 parts; then one turn of the disk will raise or depress the end of the screw $\frac{1}{2}$ mm., and to turn the disk through one division of its scale will move the screw $\frac{1}{2} \times \frac{1}{500} = .001$ mm.

To measure the thickness of the mica plate, place the spherometer on the glass plane, with the mica under the fourth leg. Turn the screw till the instrument starts to turn on the plate from the friction of the screw-leg on the plate. To the reading on the straight scale, add that on the disk, and the thickness is obtained. Several determinations should be made, and their average adopted as the true thickness, making a tabulated record of the work as directed in the previous exercises.

In some forms of the instrument, the screw-leg terminates below, in a bent lever very delicately pivoted, the long arm serving as a pointer. By watching the pointer, the exact moment of contact with the surface is known.

Instead of adjusting the zero of the instrument, it is probably better to take the mean of several readings, when standing on the plane, without the mica, and then subtract this from the mean of as many readings taken with the mica under the screw-leg. Should the reading in the first case be below zero, and in the second case above, the two must be added.

12. Exercise. — Measure the diameter of a sphere, as a croquet-ball.

First Method. — Place the sphere between two square-cut blocks whose faces press against a third block, and measure

THE PROPERTIES OF MATTER. 13

the distance between the first two by means of a good scale. To avoid errors, the angles of the blocks must be square, and the thickness of the blocks must not be less than half the diameter of the ball. Why?

Second Method. — Employ the **Outside Caliper** (Fig. 11) and a good scale. This caliper is a kind of compass with curved legs. By means of a screw, the distance between the inner faces of the points can be set so that they just glide over the

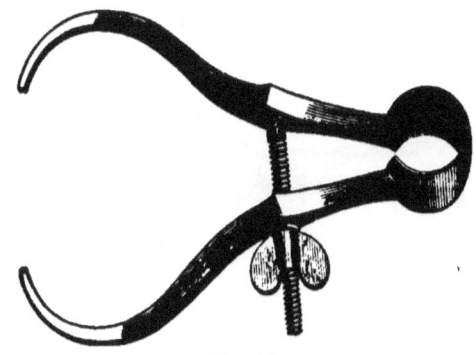

FIG. 11.

surface of the sphere. Now apply the points to the scale to determine the distance between them. This will be the diameter of the ball. Measure the ball in several different positions, record the results as in previous exercises, and obtain the mean.

This instrument can be used in all kinds of end measurements where actual contact between the instrument and the object is possible.

13. Exercise. — Measure the depth and also the diameter of a cylindrical jar, compute its contents, and compare the result with that obtained by the use of a graduated measure. Employ both English and French measure.

14   PRACTICAL PHYSICS.

To find the depth of the jar, stand vertically in it a linear scale, taking the reading on it at the point opposite the lower edge of a straight-edge resting across the top of the jar. This measurement should be made at several places, as the bottom of the jar is not perfectly level. The inside diameter can best be found by means of the **Inside Caliper** (Fig. 12). Place the instrument within the jar, and open it as far as possible. Apply the points to a linear scale, and find the distance between their outer faces. The figure shows a form of the caliper that can be used for both *inside* and *outside* measurements.

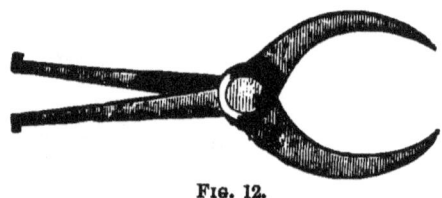

FIG. 12.

To compute the volume, apply the formula $V = \pi R^2 H$, in which $\pi = 3.1416$, $R =$ radius, and $H =$ depth.

In the absence of an inside caliper bend a piece of annealed wire into a U-shape, and use it in the same way as a caliper. The lack of elasticity will cause the points to stay wherever placed.

To measure the volume of the jar, ascertain how much water it takes to fill it, as measured both in and out with a **Graduate** (Fig. 254.) In filling the graduate, place it on a horizontal table, and in taking the reading hold the eye on a level with the surface of the water.

A simple method of finding the volume of small vessels of any shape is to subtract their weight when empty from that when filled with ice-water. This difference, in grammes, will be the volume in cubic centimetres.

THE PROPERTIES OF MATTER. 15

Record the results after some such plan as the following:—

Problem. — Determination of the contents of a cylindrical jar. Sept. 7, 1888.

Method. — Linear scale, inside caliper, water, and a graduate.

Results. —

	Depth.		Diameter.		Capacity.	
First measurement	......cm.	......in.	......cm.	......in.	......ccm.	......oz.
Second "	......	......	......			
Third "	......	......	......	......	......	......
Mean	......		......	......	......	......
Computed volume			......ccm.			......oz.
Measured "	......					
Mean	......					

14. Exercise. — Copy a scale on glass, brass, or paper.

Thoroughly clean the piece of glass or brass, and coat it evenly with a thin film of beeswax or paraffine by plunging it into a dish filled with the melted substance. Fasten both the

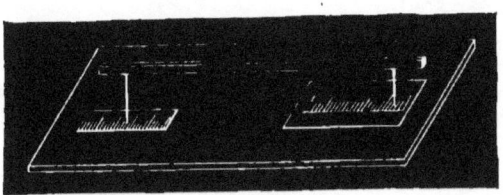

FIG. 13.

scale to be copied and the piece of brass or glass to a board by means of tacks so as to be about 50 cm. apart. In a wooden bar about 52 cm. long, 2 cm. wide, and 1 cm. thick, insert at each end, and perpendicular to the length, a stout needle-point 2 cm. long (Fig. 13). Place one of the needle-points of the bar on the first division of the scale, and with the other point draw a line in the wax cutting entirely through it. Next place the needle-point on the second division of the scale, and so on till a scale of sufficient length has been obtained. Every fifth division should be made a little longer than the others, and every tenth should be longer still. With a sharp-pointed instrument add the proper figures. If the scale is on brass, flow the plate with nitric acid for a few moments, and then wash in water. If the scale is on glass, place the plate, writing downward, over a rectangular dish made of sheet-lead, and containing a spoonful of powdered fluor-spar well moistened with sulphuric acid. Warm the vessel slightly over a lamp, being careful not to melt the wax, and set aside for a few hours. Remove the wax with a cloth wet with benzine. A scale will be found etched in the glass or brass, as the case may be.

To copy a scale on paper, substitute for one of the needles on the bar a draughtsman's right-line pen or a hard lead-pencil, inserting it in a hole bored through one end of the bar.

15. Exercise. — Test the accuracy of the divisions of a common linear scale.

To obtain the deviation from a standard scale, employ the **Beam-Compass** (Fig. 14). It consists of a graduated bar of wood or metal, the standard scale, to which are attached by clamps two points, one of which has a fine adjustment by means of a tangent-screw, C, causing the index-mark on B to move in front of a fine scale. The instrument may also be used in measuring distances between points, in copying scales, etc.

Fasten the scale to be tested to a board. Slide the point A along the bar till the distance between A and B is, roughly, equal to the distance on the scale to be tested. Then clamp A, place the point on the scale, and move B, by means of C, till the points fall exactly on the scale divisions. Use a magni-

THE PROPERTIES OF MATTER. 17

fying-glass to insure accuracy in the coincidence of the points with the divisions of the scale. Now read the distance on the beam, and combine it with the reading on C, for the distance between the scale divisions. Proceed in this way till as much of the scale as needed has been tested.

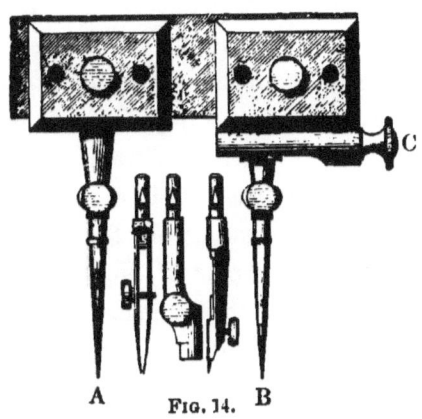

Fig. 14.

The results may be recorded as follows: —

Problem. — Testing the accuracy of a linear scale. Sept. 8, 1888.
Method. — The beam-compass.
Results. —

Scale Reading.	Compass-Bar Reading.	Scale C Reading.	True Distance.
....	....	....	....
....	....	....	....
....	....	....	....

16. Exercise. — Divide a straight line into a number of equal parts.

First Method. — Through one end of the line to be divided, draw an indefinite straight line making any angle between 30° and 45°. Lay off on this line with the dividers as many equal distances as there are parts required of the given line, beginning at the vertex of the angle. Join the last division with the free end of the given line, and through the other points of division draw lines parallel to this line. This set of parallels will divide the line as required.

To draw parallel lines, place one side of a wooden triangle, such as draughtsmen use, firmly against the edge of a ruler resting on the paper. Then on sliding the triangle along the ruler, either of the other sides will move parallel to its first position.

Second Method. — Employ a pair of **Proportional Dividers** (Fig. 15). This instrument is a variety of a double compass with a movable axis, a scale on the side indicating the ratio of the distance between one pair of points to that between the other pair as the compass is opened. It is sometimes furnished with a micrometer adjustment. The same instrument is also graduated so as to give the length of the side of any regular polygon for any given radius.

To divide the given line, set the index at that division on the scale of the instrument indicating the required number of parts into which the line is to be divided. Now open the dividers till the distance between the points of the longer legs is the length of the line; then the distance between the other two points is the required part of the line. Lay this distance off on the line, and the required division is obtained.

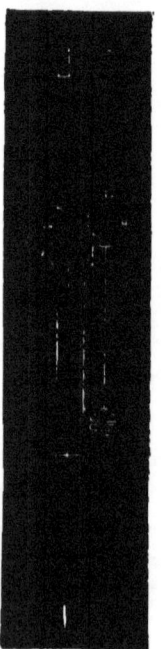

Fig. 15.

By reversing the process, a line may be extended to any number of times its original length.

17. Exercise. — Draw a circle on cardboard, and divide it into degrees.

This will require the use of a **Protractor** (Fig. 16), a brass, German-silver, or translucent horn circle or semicircle divided to degrees or half degrees. It is frequently provided with an arm turning about its centre, and so constructed that one edge of it moves exactly as a radius of the circle. For very accurate work the readings are made by means of a vernier. When not provided with an arm, a ruler will have to be used instead.

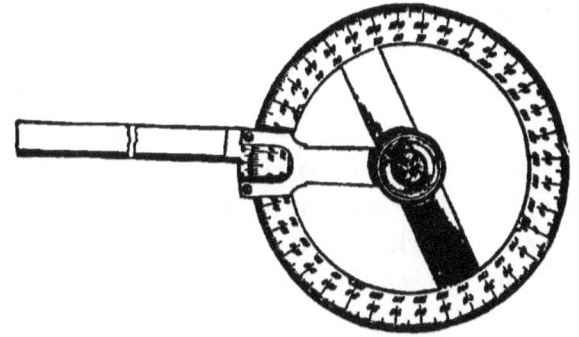

FIG. 16.

Place the protractor on the cardboard circle so that the centres coincide. Then set the radial arm at zero, and with a sharp pencil draw a line along the edge cutting the circumference of the circle. Now move the arm to the successive divisions of the scale, and mark as before. Make every fifth division line longer than the others, and every tenth longer still. Should the circle be smaller than the protractor, rule the radial lines as before; then remove the instrument, and by the aid of a ruler extend the lines toward the centre till they intersect the circumference.

18. Exercise. — Draw a triangle on a piece of cardboard, measure its angles and sides, and also compute its area.

The length of the sides can be obtained by the method of Art. 2. To measure an angle, place the protractor so that its centre is exactly on the vertex, and the zero is in one side of the angle, or that side produced. Now move the arm till it is in the other side; the reading will be the value of the angle.

The area can be computed by the aid of the formula $\sqrt{p(p-a)(p-b)(p-c)}$, in which p is the semi-perimeter, and a, b, and c are the sides. An approximation to the area of the triangle, and in fact of any plane figure, can be obtained by transferring it to what is known as cross-section paper, paper divided accurately into small squares of known size, and counting the number of these squares that the figure includes. Wherever but part of a square falls within the triangle, the value of that part will have to be carefully estimated.

Another simple method is to cut the figure out of cardboard, and compare its weight with that of a square inch or a square centimetre cut from the same cardboard.

19. Exercise. — Draw on paper a figure bounded by several straight lines, and determine its area.

Divide the figure into triangles, and compute their areas as in Art. 18. Check the work by applying the approximative methods suggested in the same article.

20. Exercise. — Find the volume of any irregular piece of stone or metal.

First Method. — Measure the volume of water the substance displaces. To do this readily and accurately, a **Cylindrical Graduate** (Fig. 254) and an **Erdmann's Float** (Fig. 17) are needed. The float is a short glass cylinder, closed at both ends, and weighted with mercury so as to cause it to take an upright position in the water. A mark is etched around the cylinder, and the position of this mark with reference to the scale on the graduate is the reading to be used. To make a float, select a

THE PROPERTIES OF MATTER.

short piece of thin glass tubing, close one end by heating it in a gas-flame, then introduce mercury or shot sufficient to cause the tube to float in a vertical position in water. Now close the other end of the tube by softening the glass and drawing it out to form a hook. A fine thread tied around the tube at a point which will be below the surface when the instrument floats in water will serve as an index. The office of the float is to avoid the error due to the liquid's climbing up the side of the cylinder, rendering it difficult to tell where the upper surface is.

Fig. 17.

Fill the graduate part full of water, introduce the float by means of a wire hook, and observe the position of the line on the float with reference to the scale on the graduate. By introducing water drop by drop, this reading can always be made a whole number. In taking readings, always read to that division on the scale that is in line with two opposite points of the line around the float. Now remove the float, and introduce the object to be measured; then replace the float, and take the readings as before. The difference between the two readings is the volume. Make several measurements, varying the quantity of water, and record the results after some such plan as the following: —

Problem. — Determination of the volume of an irregular piece of stone. Sept. 10, 1888.

Method. — Water, Erdmann's float, and graduated cylinder.

Results. —

	Float-reading before introducing the Stone.	Float-reading after introducing the Stone.
First observation	.... ccm.	.... ccm.
Second "	....	....
Third "	....	....
Mean	....	....
Volume	.... ccm.	

Second Method. — The volume of a substance whose density is known can be found in cubic centimetres by dividing its weight in grammes by the density. Consult Art. 187 and Table II.

21. Exercise. — Find the volume of a substance soluble in water, as rock-candy, common salt, etc.

Sugar and saltpetre are insoluble in strong alcohol; common salt in sulphuric ether, etc.

22. Exercise. — Find the capacity of a vessel whose form is that of the frustum of a cone.

If the vessel is a small one its capacity can be readily obtained by means of a graduated measure. If, however, the vessel is large, like a water-pail, its capacity is easily computed by the formula $\frac{\pi H}{3}(R^2 + r^2 + Rr)$, in which $\pi = 3.1416$, $H =$ depth, $R =$ radius of the top, and $r =$ radius of the bottom. These dimensions are found as directed in Art. 13.

Record the results as follows: —

Problem. — Determination of the capacity of a vessel whose form is that of the frustum of a cone. Sept. 10, 1888.

Method. — By calipers and scale.

Results. —

	Diameter of Top.	Diameter of Bottom.	Depth.
First measurement	_____cm.	_____cm.	_____cm.
Second "	....	....	....
Third "	_____	_____	_____
Mean	....	....	....
Volume			_____ccm.

THE PROPERTIES OF MATTER. 23

II. ESTIMATION OF MASS.

23. Apparatus. — As the mass of a substance — that is, the quantity of matter in it — is estimated in terms of some quantity of matter assumed as a standard, as the gramme or avoirdupois pound, there is therefore needed for its determination a copy of the standard unit, together with multiples and sub-multiples, as well as some form of balance for making the comparison.

Fig. 18.

The Balance (Fig. 18) consists of a metal beam supported on an axis perpendicular to its length, and about which it is free to turn in a vertical plane. Pans are attached at the extremities of the beam in such a manner as to move freely about an axis parallel to the axis supporting the beam. All the bearings are usually knife-edges of hardened steel resting on polished agate or steel plates. Attached to the centre of the beam is a long pointer moving in front of a graduated scale, to tell the user when the beam is horizontal. In order that the beam may not rest on the bearings when the balance is not in use, a provision is made, such that by turning a mill-head in front, the beam is lifted from its supports. To protect from air-currents when weighing, a glass case covers the

Fig. 19.

balance. This case is supported on levelling-screws to facilitate the adjustment of the instrument to that position in which the pointer is opposite the middle division of the scale. Fig. 19 illustrates a cheaper form of balance, which will be found sufficiently accurate for all requirements of this book. In fact, the common hand-balance so largely used by jewellers, if suspended from some simple support, will give, with care and patience, very good results.

Jolly's Balance (Fig. 20) is a cheap and efficient substitute for the beam balance. To construct one, proceed as follows: —

Provide a stout supporting-rod set into a square block for a base. Suspend, from an arm projecting from the top, a delicate spiral spring 30 cm. long. Two or three of these springs should be provided, varying in stiffness. They are made by winding spring brass wire closely around a cylinder 2 cm. in diameter. Wires ranging from No. 20 to No. 30 may be used, according to the sensitiveness required. Attach two small scale-pans as seen in the figure, the lower to be used in density determinations for weighing objects in water, an adjustable shelf being necessary to support the tumbler. On the standard back of the spring, tack a long strip of mirror on which is a millimetre scale (see Art. 14). Instead of etching the scale on the face of the mirror, it will be simpler to cut it in the amalgam, and then cover the back with black paper. A paper scale cemented along the edge of the mirror will answer nearly as well. Just above the scale-pan, attach a small white bead, or a wire pointer, at right angles to the spring, to serve as an index. The scale-reading is where the line joining the index with its reflection intersects the scale.

The **Weights** (Fig. 21), when large, are made of brass; when small, of either aluminum or platinum. Each weight has its value stamped upon it, and should have its assigned place in a neat box. In handling them, always use the small pair of pincers found in the case. The milligramme weights are not much used where the balance-beam is graduated. Their place is supplied by what is known as the **Rider**, a U-shaped piece of wire, usually weighing one centigramme, that is placed astride of the beam. If the arms, for example, are each divided into ten parts, then the rider placed at division six represents a weight of six-tenths of a centigramme, or 6 milligrammes, placed on the scale-pan. By estimating fractions of these divisions, or using lighter riders, fractions of milligrammes can be weighed.

Fig. 21.

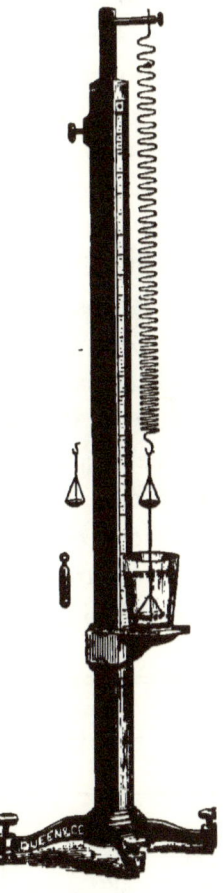

Fig. 20.

24. Exercise. — Find the weight of a small piece of any substance.

Set the balance vibrating, with empty pans, to determine the position of the resting-point of the pointer on the ivory scale. Since it will take too much time to allow the beam to come to rest, it is better to determine where the resting-point would be by observations made on the oscillations. This would be an easy matter if the resting-point was the middle point of the arc described by the pointer. But the distance the pointer swings each side of the resting-point gradually diminishes; hence it will be necessary to average a number of oscillations after the manner described below.

Suppose the ivory scale to be divided into thirty parts. To avoid the use of signs in distinguishing the deviations of the pointer to the right or left of the middle division, number the scale from left to right, the extreme left being marked 0, the middle 15, and the extreme right 30. Then record five vibrations of the pointer. This will give the greatest and the least arc to the side toward which the pointer was first observed to move. These would be recorded as follows: —

Turning-Points.				Resting-Point.
Left.		Right.		
First vibration,	3	Second vibration,	25	
Third "	5.3	Fourth "	22.2	$\text{Mean} = \dfrac{5.1 + 23.6}{2} = 14.35.$
Fifth "	7			
Mean . .	5.1	Mean . . .	23.6	

Hence 14.35 is the resting-point of the unloaded balance. By means of the levelling-screws, the balance could be adjusted

so that the resting-point would be the middle division of the scale, but it is unnecessary. Now place the article to be weighed on the left-hand scale-pan, and as near the middle as practicable. In the other pan, place a weight such as you estimate will balance it. Now throw the balance in action, and watch the pointer to determine which pan is the heavier, since, if the equilibrium is not close, the pointer will move sharply to one side of the resting-point, whereas, if close, the resting-point will be seen to be near the middle of the arc described. If the weight is found too heavy, remove it, and substitute a smaller one; if too small, a heavier one. Let us suppose that the twenty-gramme weight is found too heavy. Then replace it by the ten-gramme weight. If this is found too light, then add the five-gramme weight; if still too light, add two more; if now too heavy, then replace the two-gramme by the one-gramme; if now too light, add .5 gramme; if still too light, add .2 gramme; if now too heavy, replace .2 gramme by .1 gramme, and so on. When traced down to centigrammes, the centigramme rider may be used if the balance has a graduated beam, and the milligrammes determined.

When the swinging of the pointer shows that the balance is nearly secured, find the resting-point of the loaded balance in the same manner as in the case of the unloaded. Let us suppose that is found to be 13.5. This shows that the right-hand pan is the heavier.[1] To determine the amount, place .001 gramme in the left-hand pan, and again determine the resting-point. Let us suppose that this resting-point is found to be 15. Hence it follows that .001 gramme has moved the pointer through 15 − 13.5 = 1.5 divisions, whereas it was

---

[1] This applies to a balance having the ivory scale below the beam. Where this scale is above the beam, the opposite is indicated.

necessary to move it $14.35 - 13.5 = .85$ division. Therefore the amount to be placed in the left-hand pan, that is, the amount to be deducted from the weights in the right-hand pan, is $(.85 \div 1.5) \times .001 = .00057$ gramme. Finally add together the weights in the pan, and deduct from the sum .00057 gramme to obtain the weight of the substance.

To prevent errors in determining the total value of the weights on the pan, first, add up the values of those which the empty pockets of the case show to be in use; secondly, check the same by adding the values of the weights as they are removed from the pan to the case.

In weighing with fine balances, observe the following precautions : —

1. Remove all dust from the pans by means of a camel-hair brush.

2. Set the balance in action by turning the mill-head to find the resting-point of the pointer, as it is not always the middle division of the scale.

3. When it is necessary to stop the swinging of the beam, wait till the pointer is directly over the centre of the scale, then turn the mill-head. To do otherwise jars the instrument, and injures the bearings.

4. Always stop the swinging of the balance when adding or removing weights.

5. The position of the observer should be central so as to avoid any parallax in reading the scale.

6. Always put on weights in the order in which they come in the box or case, handling them with the pincers, and never with the fingers.

7. When the weighing is finished, replace the weights in the box, and in no case leave the balance swinging if there is any provision for avoiding it.

THE PROPERTIES OF MATTER.

8. Avoid all air-currents, and have a firm support for the balance, so that there may be as little trembling as possible.

9. Never be in a hurry, as accurate weighing is slow work.

10. All substances liable to injure the pans, when being weighed, must be placed in appropriate vessels; a watch-glass of known weight will be found very useful for such purposes.

To weigh with Jolly's balance, place the substance in the upper pan, and take the reading of the index. Now remove the substance, and add weights till the same reading is obtained. The sum of the weights is evidently the weight of the substance. If, with the weights given, it is found impossible to bring the index to the same reading, add weights till a reading less than the required one is obtained, but such that the smallest weight on hand, when added, brings the index too low. If, for example, the reading required is 300, and 1 mg. changes it from 299 to 301.5, then the weight required will be $\frac{1}{2.5}$ of 1 mg., or .4 mg.

25. Exercise. — Determine the weight of a number of lead balls by weighing five of them separately, selected at random, finding their average, and multiplying that average by the number of balls in the lot. Weigh each ball several times.

The record may have the following form: —

Problem. — Determination of the weight of a number of lead balls.

Method. — Weighing five of them separately, selected at random, finding their average, and multiplying that average by the number of balls in the lot.

Results.—

	First Ball.	Second Ball.	Third Ball.	Fourth Ball.	Fifth Ball.
First weighing	....		....		
Second "	....	....	....	....	
Third "	....	....	....	....	....
Mean . . .	....		....		....

Average weight of a ball = ... Weight of lot = ...

26. Exercise. — Determine how many grains of wheat in a bushel, 60 pounds, by weighing five lots of 20 grains each, and from these data determining the average weight of a grain.

Devise a convenient form for recording the results.

27. Exercise. — Determine how much chalk it takes to write your name on a common blackboard. Also determine the amount used in writing your name on a slate. Which is the more economical surface on which to use chalk?

Use the mean of several times' writing. Devise some suitable form for tabulating results.

28. Exercise. — Determine the weight of a silver five-cent piece, and also that of a silver dollar. What is the value of the silver dollar on the standard adopted for the five-cent piece?

29. Exercise. — Determine the percentage of water in crystallized sodium sulphate.

Weigh carefully a large crystal of the salt, and then set it aside for several days in a place where dust is excluded. The crystal will fall to powder, owing to the escape of the water of crystallization. Find the weight of the powder, and then compute the percentage of loss.

THE PROPERTIES OF MATTER. 31

30. Exercise. — Compare the weights of equal volumes of several substances, expressing each in terms of the lightest taken as the unit.

Get a competent mechanic to make out of lead, brass, tin, iron, maple, lignum-vitæ, etc., cubical blocks 1.5 cm. on an edge, that is, blocks of equal volumes, and determine their weights in the regular way.

31. Exercise. — Determine the length of a twisted piece of wire.

Weigh several straight pieces of wire of the same material and gauge-number, and also measure their lengths. From these data, compute the average weight of a centimetre of the wire. Now weigh the twisted piece, and compute its length.

The results may be recorded as follows: —

Problem. — Determination of the length of a twisted piece of wire.
Method. — Weighing and measuring.
Results. —

	Length.	Weight.	Weight per Centimetre.
Wire No. 1	....	....	....
" " 2	....	....	....
Mean			....

Weight of the twisted piece, ... Computed length, ...

32. Exercise. — Cut a piece of smooth tin-foil of some simple figure; determine its area and weight. Compute its thickness, allowing that 1 ccm. of tin weighs 7.29 grammes. Compare the result with that obtained by the use of the micrometer caliper. Devise a suitable form of record in the note-book.

33. Exercise. — Make out of aluminum-foil a centigramme weight.

First find both the area and the weight of a piece of the foil, and then compute how large a piece will be required to weigh one centigramme. As the thickness of the metal may not be uniform, cut the piece a trifle larger than the estimated one. By means of a fine file, adjust the piece to the required weight.

34. Exercise. — Measure the cross-section of a capillary glass tube.

Clean the tube thoroughly by washing it, first in nitric acid, then in distilled water, then in a solution of sodium hydrate, and finally in distilled water. To do this, connect a small glass syringe to the tube to be washed, by half a metre of rubber tubing; then, by placing one end of the capillary tube in a vessel containing the wash, the corrosive liquid is made to pass backward and forward through the tube without coming in contact with the hand. After thoroughly washing the tube with distilled water, rinse it out with alcohol, and then dry it by passing it back and forth over a gas or spirit flame. Fill the tube part full of pure mercury by dipping it into a dish filled with the liquid; lay it on a horizontal surface and by means of a pair of dividers and a diagonal scale, ascertain the length of the mercury filament. Pour the mercury out of the tube into a suitable vessel of known weight, and ascertain its weight. Now the volume of mercury in the tube is equal to the weight of mercury divided by 13.6, and the cross-section of the tube equals the volume of mercury divided by the length of the filament, provided the centimetre and the gramme are the units used. How would you find the diameter?

THE PROPERTIES OF MATTER.

III. IMPENETRABILITY.

35. Apparatus. — The principal appliances needed in the study of this subject are as follows: **Battery-Jar, Cubes of Wood, Funnel, Funnel-Tube with a Stop-Cock, Rectangular Prism of Wood, Wide-Mouthed Bottle, Tumbler, Glass Tubing** and a **Cylindrical Graduate.**

36. Exercise. — Graduate a glass battery-jar of two or more litres capacity to decilitres by means of a graduated measure and water, marking the divisions on a paper strip pasted on the outside. Cut out of wood two cubes, one decimetre and one-half a decimetre on their edges respectively. Give each block a coat of thin shellac varnish, made by dissolving shellac in alcohol, or of hot paraffine. Now fill the jar about half full of water, and observe by the scale on the side exactly how much there is. Then place in the jar these blocks successively, holding the block under water by means of a slender knitting-needle, and, while submerged, note exactly the volume of water as shown by the scale on the side of the jar. Compare the increase of volume with the volume of the block. Inference.

Owing to the climbing-up of the water on the side of the jar, it is difficult to get the exact position of the surface. This may be partly corrected by rubbing a thin film of paraffine on the inside of the jar opposite the scale.

37. Exercise. — Partly close the throat of a funnel with a perforated cork. Fit a second cork accurately to a wide-mouthed bottle with the stem of the funnel passing tightly through this cork (Fig. 22). Bore a small hole through this cork of about 2 mm. diameter. Now pour water into the funnel, holding a finger firmly over this small hole in the cork. Observe the difference on removing the finger. Why would the water not run in at first? What do you learn from this experiment about water and air?

As large corks are usually very open, they should first be boiled in paraffine. In fitting corks, a flat wood-file and a small round file will be indispensable.

Fig. 22.

38. Exercise. — Make a wooden rectangular prism 5 cm. by 2 cm. by 20 cm. In a cylindrical jar, put 100 ccm. of water. Graduate the longest edge of the prism to centimetres. Insert the bar vertically into the water. Mark on the side of the jar the change of level of the water; then remove the bar, and ascertain what volume of water will produce the same change of level. How does this volume compare with that of the part of the bar submerged? Repeat the experiment two or three times, varying the amount of the bar submerged. What property does this experiment show that matter possesses?

39. Exercise. — Adjust a U-shaped delivery tube to a bottle containing a known quantity of water, the inner end of the tube opening beneath the water, and the outer end opening within a common tumbler (Fig. 23). Pass through the same cork a funnel-tube with a stop-cock. All these fittings must be made air-tight. For this reason, a rubber stopper is to be preferred. Now shut off the stop-cock, and pour into the funnel a measured quantity of water. Then turn the stop-cock, and let the water pass into the bottle. Measure the water in the

THE PROPERTIES OF MATTER. 35

tumbler, and compare the amount with that poured into the bottle. Pour in another known quantity, and then ascertain the amount in the tumbler. Continue in this way till at least five separate amounts of water have been added to the bottle. Find the average of the five amounts of water poured into the bottle, and also of the five amounts collected in the tumbler. Measure the amount of water left in the bottle, and correct these averages by the difference between this amount and the amount in the bottle at the beginning. What inference can you make from comparing these averages?

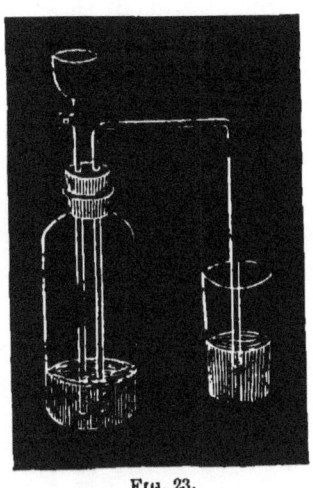

Fig. 23.

40. Exercise. — Fit a delivery-tube, and also a funnel-tube provided with a stop-cock, to a well-corked bottle (Fig. 24). Let the delivery-tube open beneath an inverted cylindrical graduate filled with water over a small pneumatic trough. Pour a known quantity of water into the funnel-tube, opening the stop-cock as soon as the air has escaped from the throat of the funnel. Measure the air now found in the graduate. Repeat the opera-

Fig. 24.

tion several times, and measure the air in the graduate after each pouring. Each of these results should be corrected by subtracting from it the capacity of the tube below the stop-cock. Why? This capacity can be easily determined by closing the end with the finger, and finding the amount of water required to fill the tube. Compare the amounts of water and air. Inference.

The funnel-tube may be replaced by a common funnel joined to a glass tube by a rubber connector. A pinch-cock (Fig. 269) on the rubber will serve for a stop-cock.

41. Exercise. — Float a cork on the surface of water. Cover it with a wide-mouthed bottle, and push the glass vessel, mouth downward, into the water. What phenomenon do you observe? Insert one arm of a U-tube made of glass tubing under the edge of the bottle, and again push it down into the water. Does the same phenomenon occur as before? What do you learn by holding a burning match over the outer end of the U-tube as the vessel is pushed into the water? What is the lesson from this experiment?

42. Exercise. — Fill a tumbler level full of water. Now drop in carefully, and one at a time, a number of small nails. In the fact that the water does not flow over the side of the vessel, have you an exception to the principle that matter is impenetrable? State the evidence on which your answer is based.

IV. DIVISIBILITY.

43. Apparatus. — The appliances needed are **Balance, Weights,** and **Micrometer Caliper.**

THE PROPERTIES OF MATTER.

44. Exercise. — Put two or three drops of nitric acid on a piece of copper; let it stand till the boiling action ceases, and then wash it off in a half-litre of clear water. To this add a few cubic centimetres of ammonia water. The blue discoloration denotes the presence of copper. Have you any evidence that copper is present in every drop of the liquid? What physical property do you find belongs to copper? Determine the amount of copper in each cubic millimetre of the blue solution by weighing the copper before applying the nitric acid, and after washing it in the water, and by measuring the solution.

45. Exercise. — Dissolve .01 gramme of an aniline dye in one litre of water. By dropping water from a bottle or a dropper into a graduated measure, ascertain, by averaging five trials, the number of drops in a cubic centimetre of the solution. Now compute how much aniline there is in each drop of the colored solution. What property does aniline possess to a very marked extent?

FIG. 25.

46. Exercise. — Weigh accurately a clean silver coin. Place on it two or three drops of nitric acid, letting it remain till the action of the acid on the silver ceases. Then wash it in a tumbler of pure water. Stir the water thoroughly, and divide it into two parts. To the one add a few drops of a strong solution of common salt, and to the other add a few centimetres of ammonia-water. Remembering that common salt turns a solution of silver milky, and ammonia-water turns a solution of copper blue, what inference can you draw from the experiment? Now weigh the coin, and compute how much silver and how much copper there is in each cubic centimetre of the solution.

NOTE. — American silver coin is 10 per cent copper.

47. Exercise. — Measure with a micrometer caliper the thickness of a piece of mica. Measure the thickness of the thinnest piece you can split from the original one. Is it due to the lack of skill, or to some property inherent in the mica, that you are unable to obtain a piece still thinner?

V. POROSITY.

48. Apparatus. — For the study of porosity, procure a **Gas-Bottle**, a **Three-Necked Bottle**, some **Glass Tubing**, a **Common Tumbler**, a **Cylindrical Graduate**, and a **Florence Flask**.

49. Exercise. — Determine whether rubber is porous or not.

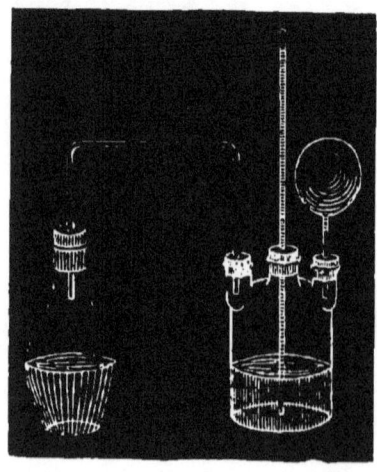

Fig. 26.

Fit a delivery-tube to a gas-bottle (Fig. 26), and let it open into a three-necked bottle in which is some water. Through the middle neck, pass a long glass tube reaching to the bottom of the bottle. Close the third neck with a perforated cork, through which passes a short piece of glass tubing. Now put some zinc clippings in the gas-bottle, and cover them with dilute sulphuric acid, one part acid to three of water. Hydrogen gas will be given

THE PROPERTIES OF MATTER. 39

off, which will pass over into the second bottle. If you press your finger firmly over the end of the tube in the third neck, you will notice that the water will rise higher and higher in the second one till it flows out of the top (why?). Remove the finger, and tie a small rubber balloon over the end of the tube. The water in the second tube soon begins to rise, but does not get beyond a certain point, notwithstanding that the evolution of gas still continues in the gas-bottle. What becomes of this gas? What must be the physical structure of the rubber?

All the corks must be firmly tied down with stout cord to prevent them from being driven out by the pressure of the gas.

50. Exercise. — Test a piece of chamois for porosity. Tie it over the end of a stout glass tube about two-thirds of a metre long, and 15 mm. in diameter. Hold the tube over a large glass vessel, and with a test-glass (Fig. 27), or some vessel with a lip, pour mercury into the tube, filling it nearly full. What does this experiment teach about the structure of chamois? Would very fine shot substituted for the mercury act in the same way? Why? What property of mercury is also revealed in this experiment?

Fig. 27.

51. Exercise. — Test thin sheets of several substances for evidences of porosity.

Select two plain tumblers of thin glass and of the same size. After thoroughly warming one of them, fill it about half full of boiling water. Now cover the glass with a piece of cardboard, and invert the other glass upon it (Fig. 28). After the lapse of one or two minutes, ascertain if there is any change in the condition of the interior of the upper tumbler. What does the experiment teach regarding the structure of cardboard?

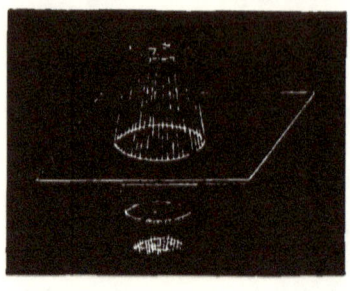
Fig. 28.

Now re-heat the water, and test in a similar manner pieces of felt, chamois, and wood. In the case of wood, ascertain in what manner a change in the thickness of the strip affects the phenomenon.

52. Exercise. — Fit to a Florence flask a delivery-tube leading to an inverted bottle standing over the pneumatic trough (Fig. 29). Fill the flask, tube, and inverted bottle with

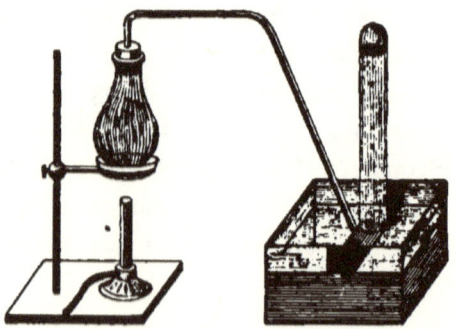
Fig. 29.

water, and apply heat to the flask. After the boiling of the water has continued for some time, remove the lamp, and examine closely the collection-bottle. Whence the gas found in it? Repeat the experiment, filling the apparatus this time

THE PROPERTIES OF MATTER. 41

with water that has been boiled for some time in an open vessel. What property does this experiment show that water possesses?

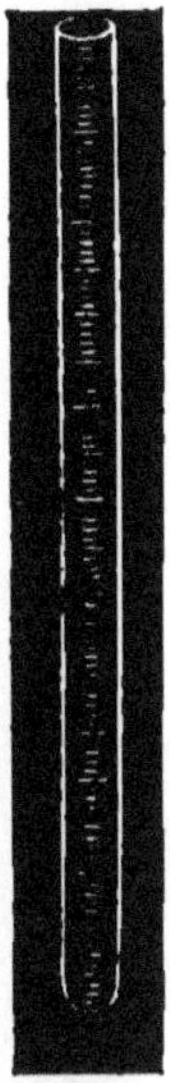

53. Exercise. — Put 75 ccm. of water in a cylindrical graduate. To this add 2 ccm. of finely powdered sugar. After the sugar has dissolved, observe the volume of the solution. Account for the shrinkage.

54. Exercise. — In a glass tube graduated to tenths of cubic centimetres (Fig. 30), put 30 ccm. of water. Then pour in very gently, tipping the tube so that it will flow down the wall of the tube, 20 ccm. of strong alcohol. Hold the finger firmly over the mouth of the tube, and shake vigorously. Observe the volume of the mixture. What is the percentage of shrinkage? What has been previously proved of water that will assist in explaining this phenomenon?

FIG. 30.

VI. INDESTRUCTIBILITY.

55. Apparatus. — The appliances needed are **Test-Tubes, Beakers, Flasks,** and a **Balance.**

56. Exercise. — Prepare a saturated solution of calcium chloride by adding the salt to 20 ccm. of water until it refuses to dissolve any more. Pour enough of the solution into a test-tube (Fig. 31) to fill it a little less than half full. In a similar manner, prepare an equal quantity of a saturated solution of sodium sulphate in a

FIG. 31.

second test-tube. Place the test-tubes in some suitable vessel to prevent them from overturning, and determine their united weight. Now pour one solution into the other, and shake quickly. Observe the change that takes place. Ascertain if the weight has changed.

Bottles may take the place of the test-tubes.

57. Exercise. — Support a piece of phosphorus of the size of a small pea on the end of a wire of the form shown in Fig. 32. Put about 400 ccm. of water colored with blue litmus into a thin beaker of about one litre capacity. In this place the wire stand supporting the phosphorus, and invert over it a Florence flask, resting its neck on the edge of the beaker closing its top, the mouth of the flask reaching nearly to the bottom of the colored water. After thoroughly drying the outside of the apparatus, place it on one of the pans of a balance, and exactly counterpoise it. Let it remain undisturbed for twenty-four hours, and then examine the flask carefully to ascertain what changes have taken place within it.

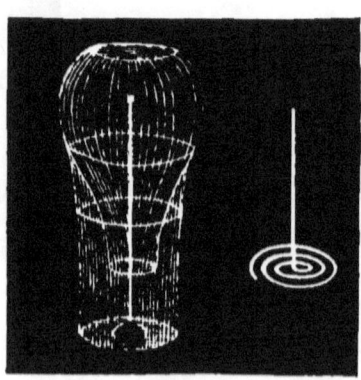

FIG. 32.

Would the same weight be required to counterpoise the apparatus if no phosphorus had been placed on the wire support? Has the weight of the beaker and its contents been diminished by the disappearance of the phosphorus? Have the phosphorus and some of the air in the flask been destroyed?

THE PROPERTIES OF MATTER. 43

To prevent the evaporation of water from the beaker, the edge should be coated with paraffine.

58. Exercise. — Fill a test-tube, say 2.5 cm. diameter and 15 cm. long, with water; cover the mouth with the thumb, and invert the tube in a small beaker partly filled with water (Fig. 33). The tube should be full of water, no air having been admitted in inverting it. Slip over the tube a short piece of large glass tubing, as a piece of lamp-chimney, to support the test-tube in a vertical position. Place a small piece of zinc, say .05 gramme, beneath the mouth of the tube. Pour into a small beaker a few centimetres of strong sulphuric acid. Place the vessel supporting the tube, and also the beaker of acid, on the pan of a fairly good balance, and counterpoise them. Pour the acid into the vessel supporting the tube, replacing the beaker in the scale-pan. The acid will act on the zinc, and hydrogen gas will be seen to collect in the top of the tube. From time to time, as these changes proceed, set the balance in action to ascertain if there is any gain or loss of matter. Inference.

FIG. 33.

Caution. — If too much zinc is used, the tube will not hold all the gas evolved.

59. Exercise. — Select two thin glass beakers, each having a capacity of about 100 ccm. Into one put 50 ccm. of a solution of lead nitrate, and into the other put 25 ccm. of a solution of potassium chromate. Place the two beakers on the pan of a balance, and counterpoise them. Now pour the contents of one beaker into the other, restoring the beaker to its place on the scale-pan. Observe the changes which occur, and determine if they are attended with any gain or loss of matter.

VII. COHESION.

60. Apparatus. — Bar of Lead, Balance, Disks of different substances, Funnel with Stop-Cock, Glass Bulb, Spring-Balance, Crucible, Evaporating Dish, etc.

61. Exercise. — Bore a hole, about 2.5 cm. in diameter, in a block of wood, and, using it as a mould, cast two disks of lead 2.5 cm. thick. Dress one surface of each disk to a plane. Now press the two disks firmly together, giving one of them a slight twisting motion. In this way, one piece can be made to hold up the other. What force is brought into action? What do the facts that pressure, as well as a very smooth surface, is necessary to the success of this experiment, teach regarding this force?

62. Exercise. — Measure the cohesion of paper or wire.

Cut a rectangular piece of the paper to be tested 25 cm. long and 10 cm. wide. Fold over each end, fastening it with glue, forming a loop or hem. In these insert stout wooden rods somewhat longer than the width of the paper. Connect the ends of one of these rods to the hook of a spring-balance by means of a wire or cord bail. Fasten the other rod to some suitable support. Now pull steadily on the ring of the balance, recording the reading observed at the moment the paper is parted.

Wire can be tested in a similar manner by fastening one end to some firm object, and the other end to the hook of the balance. Compare strength with the cross-section. Ascertain if length affects the strength.

63. Exercise. — Compare the cohesion of water, alcohol, glycerine, etc., by ascertaining the degree to which such cohesion affects the size of falling drops of the liquid.

Support by a clamp a funnel provided with a stop-cock so that its stem is about 1 cm. above the surface of a glass sphere. A small Florence flask with a round bottom, or the bulb of a common air thermometer, may be used for the sphere. Pour some of the liquid to be tested in the funnel, and open the stop-cock so that the liquid will flow at such a rate upon the sphere as to drop from the under surface at the rate of two drops per second. Now catch one hundred of these drops in a beaker, and determine their weight. Make at least three determinations for each liquid.

Repeat the experiment, changing the drop-rate to one in two seconds. How is the size of the drop affected?

Ascertain the effect on the drop-size of employing a smaller glass sphere.

Record the results as follows: —

COHESION.

Problem. — Measurements of drop-size of several liquids. Sept. 17, 1888.

Method. — Weighing one hundred drops of liquid.

Results.—

	Drop-Rate, ...			Drop-Rate, ...		
	Water.	Alcohol.	Glycerine.	Water.	Alcohol.	Glycerine.
First trial						
Second "						
Third "						
Mean						
Weight of drop						

64. Exercise. — Measure the force necessary to pull a disk away from a liquid.

Remove one of the scale-pans from a balance, — the jeweller's form answers the purpose nicely, — and suspend in its

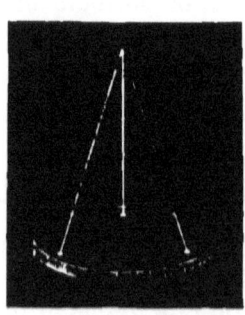

Fig. 34.

place a glass disk, about 5 cm. diameter, by means of three threads attached at points 120 degrees apart (Fig. 34). After accurately counterpoising the disk, place a vessel of water below it, raising it till the surface of the water touches the under surface of the disk, being careful to keep the beam horizontal. Now add weights to the scale-pan till the disk is pulled away from the water. These weights must not be dropped into the pan, as the sudden jar would tend to separate the disk from the water. Ascertain whether you have pulled a column of water apart, or pulled the plate away from the water. Measure the disk, and then compute the force per square centimetre required to effect the separation.

The disk must be perfectly clean, and in bringing it in contact with the liquid, first touch one edge, and then gradually shut down the disk upon the liquid, thus excluding all air

THE PROPERTIES OF MATTER.   47

bubbles, and insuring good contact. The threads can be attached to the disk by means of tough sealing-wax. See Art. 607.

Try other liquids, as mercury, alcohol, glycerine, etc.

Try successively disks made of brass, tin, zinc, etc.

Amalgamate these disks with mercury before measuring the force required to separate them from mercury.

Measure the force required to separate from water a glass disk coated with a film of sweet oil.

Construct a comparative table of all the results.

65. Exercise. — Select a soft glass tube about 25 cm. long and 2 cm. in diameter. Close one end in the flame of a blow-pipe, and then bend the tube to a V shape with its branches widely diverging, and the closed arm 3 cm. the longer. Nearly fill the tube with water, and boil it uniformly till nearly a quarter has boiled away. Remove the flame, and close the open end air-tight with a rubber stopper. When cold, hold the hand against the end of the long arm; let the water in the tube fall against it. The water will be found to remain suspended in that arm on holding it in a vertical position, instead of falling back to the level of the water in the other arm, requiring quite a jar to break it away from the end of the tube. Explain.

66. Exercise. — Dissolve 100 grammes of powdered alum in half a litre of hot water. Hang in the solution strings, twigs of plants, or wire forms, and set aside for twelve hours. Make a careful study of the alum crystal. Diagram one. Copper sulphate may be substituted for the alum. Make a mixture of the two solutions in an evaporating dish or saucer, and set aside till crystallization occurs.

67. Exercise. — Wet the surface of a strip of glass with a solution of ammonium chloride. As it begins to dry, examine it carefully under a microscope. A good botanizing-glass will answer the purpose.

68. Exercise. — Melt a quantity of sulphur in a Hessian crucible or common tea-cup. The vessel should be at least two-thirds full. As soon as the sulphur is melted, set it aside to cool, leaving it till a thin crust forms over the top. Now break through the crust, and pour out the liquid interior. Examine with a common magnifier the interior of the cavity, determining the form of the crystals.

69. Exercise. — Prepare a saturated solution of common salt. Pour the solution into a common saucer or an evaporating dish, and set it aside protected from dust. In a few days, the bottom of the dish will be covered with crystals curiously grouped. Make a close study of the shape of the crystals, and the manner of grouping.

By observing the following directions, quite large salt crystals can be obtained: —

" The salt to be crystallized is to be dissolved in water, and evaporated to such a consistency that it shall crystallize on cooling. Set it by, and when quite cold, pour the liquid part from the mass of crystals at the bottom into a flat-bottomed vessel. Solitary crystals will form at some distance from each other, and gradually increase in size. Pick out the most regular, put them into another flat-bottomed vessel a little apart from each other, and pour over them a quantity of fresh solution of the salt evaporated till it crystallizes on cooling. Alter the position of every crystal once at least every day with a glass rod, that all the faces may be alternately exposed to the action of the liquid, for the face on which the crystal rests never receives any increase. By this process, the crystals will gradually augment in size. When they have acquired such a magnitude that their forms can easily be distinguished, the most regular are to be chosen, or those which have the exact shape which you wish to obtain. Each of them should be put separately into a vessel filled with a portion of the same liquid, and turned by the glass rod several times a day. Whenever it is observed that the angles and edges of the crystals become blunted, the liquid must immediately be poured off, and fresh liquid put in its place; otherwise the crystals will be infallibly destroyed."

70. Exercise. — Procure pieces of mica, Iceland spar, roll sulphur, common white chalk, alum, coal, feldspar, blue vitriol, sal ammoniac, galena, pyrites, etc. With a knife, try to split these substances in different directions. What do you discover with respect to the relative ease of breaking or splitting each of these substances in different directions? Study closely the surfaces exposed at each separation, and see if it indicates the breaking-up of a structure, or the splitting-apart of two or more complete structures. Examine the corners; in short, note every thing having any bearing on the molecular structure of the substance.

71. Exercise. — Classify a number of substances with reference to their hardness on Mohr's scale.

Mohr's scale is the hardness of the following ten substances, which are ranked as shown by the attached numbers: 1, talc; 2, gypsum; 2.5, mica; 3, calcite; 4, fluor-spar; 5, apatite; 5.5, scapolite; 6, feldspar; 7, quartz; 8, topaz; 9, sapphire; 10, diamond.

To determine the hardness of any substance, draw a file over it with considerable pressure, and observe whether the depth of the cut made in the specimen is greater, less, or equal to, that made in some one of the minerals of the scale. If, for instance, the cut made in the substance is less than that made in 6, and deeper than that made by the same pressure of the file in 7, it ranks in hardness between 6 and 7.

Glass, slate, marble, gypsum, galena, hematite, magnetite, fluor-spar, copper, silver, steel, etc., are some of the substances easily obtained for the purposes of this experiment.

VIII. ELASTICITY.

72. Apparatus. — The appliances required for studying elasticity are such as any one accustomed to the use of tools can readily make by following the directions given. A stock of wires, wooden bars, and metal and wooden rods of various sizes and numbers, will be needed for testing.

73. Exercise. — Study the elasticity of solids when manifested by pressure.

Procure several balls, one each, of wood, glass, ivory, etc., each having a diameter of about 2 cm. Drop them from the same height on a marble slab, and compare the heights to which they rebound. Double the height, and repeat the comparison. Triple the height, and compare. Now repeat the experiment, having coated the slab with a thin film of paste made of olive-oil and common whiting thoroughly mixed together. Compare the marks made in the film by the balls when merely laid on the marble, with those made when the balls fall from a height. What inference follows respecting the various substances of which the balls are made?

THE PROPERTIES OF MATTER. 51

74. Exercise. — Show that the amount of extension of a wire is proportional to its length, and to the weight which produces the extension, within certain limits, and is inversely as the cross-section.

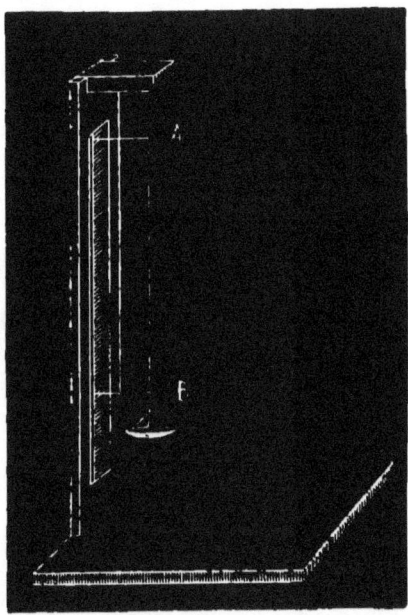

FIG. 35.

Construct a stout wooden frame of the form shown in Fig. 35, having the upright at least one metre high. Attach to it a millimetre scale. Suspend in succession *straight* pieces of iron, steel, copper, brass, etc. wires of the same size, from a hook on the under side of the arm at the top of the vertical support, fastening to the lower end a scale-pan made of tin or brass. Fasten to the wire two pointers, A and B, and record the difference in the readings for the length AB. Now put some known weights in the pan, and again determine AB. Add more weights, and again read the distance AB. Continue doing this, if possible, till the wire breaks. After each reading is taken, remove all the weights, and record the length of the wire. Tabulate the results. Compare the increase in length each time with the total weight in the pan, and you should find that the wire has increased in length proportionally to its length, and to the load in the pan, up to a certain point; also, that the wire has returned each time to its original length up to the same point, and that beyond that point, known as the *limit of elasticity*, the elongation is not uniform.

Repeat the experiments, employing a wire of somewhat larger cross-section, and make the same changes in the weights. Compare the effects on length with those previously determined. Inference.

Construct a curve from the data obtained, and point out how it proves the law. For method of constructing curve, see Art. 598.

75. Exercise. — Determine the laws governing the deflection of beams.

Construct of wood an apparatus such as shown in Fig. 36. A and B are pyramidal pieces of hard wood 20 cm. high, standing on the base N, the distance between their upper edges being determined by means of the linear scale S. M is a bar, wood or metal, 1 metre long, 2.5 cm. wide, and 5 cm. thick, whose deflection is to be measured. C is a small clevis made of sheet brass, placed exactly over the centre of the bar, supporting the scale-pan F. RLE is a bent lever made of wire, the weight of which is sufficient to keep its foot in close contact with C as the bar M deflects. Narrow grooves filed in the wire on top of the post H will make it easy to secure it in place by means of wire staples. Freedom of motion must be secured with as little play as possible. K is a scale of equal parts.

Record the reading of the pointer, then keep adding weights and recording the reading till a sufficient number have been made for comparison. Compare the amount of deflection with the weights applied.

Try a bar of the same material, having double the breadth, and the same thickness as the first. Inference. Try one

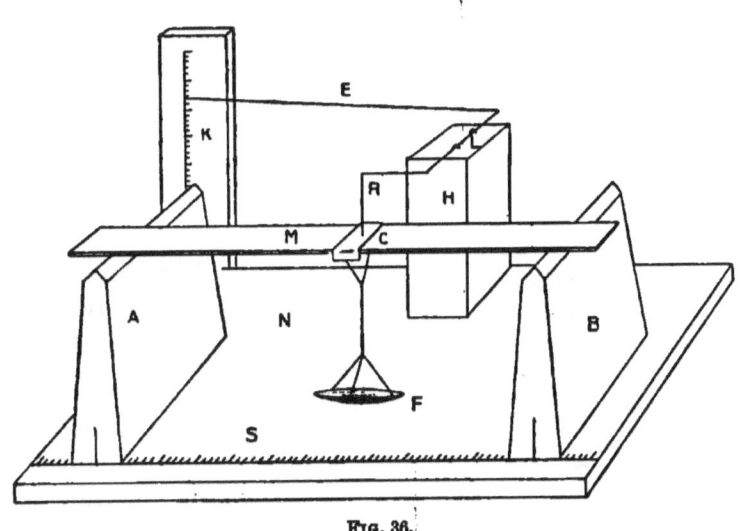

Fig. 36.

having double the depth, and the same width as the first. Inference. Try one having double the length, but the same width and thickness as the first. Inference.

Construct a curve from the data.

76. Exercise. — Determine the laws of elasticity by torsion.

Construct of wood an apparatus such as is shown in Fig. 37, making L 90 cm. long. W is the wire to be experimented on, the upper end being squared, and fitted snugly into a metal plate A to keep it from turning, the lower end passing through the pulley E, and entering a round hole in the base M. The pulley is made with a hub, so that a set screw can be used to clamp it rigidly to the wire. F is a pointer moving over the arc D. H and K are common iron pulleys, over which cords from the pulley E draw so that weights placed in the pans twist the wire or rod. The shelves B and C are placed so that the spaces AB, BC, and CD are each equal to 30 cm. Observe the reading of each pointer, then place equal weights in the pans, and observe the change in the readings. Compare the change at B with that at C and at D. How does length affect the amount of twist? Increase the weights in the pans, and determine, from the change in the readings, how the force applied affects the twist. Substitute for W a rod of the same material having a different diameter. Use the same weights, and determine how a change in diameter affects the amount of twist.

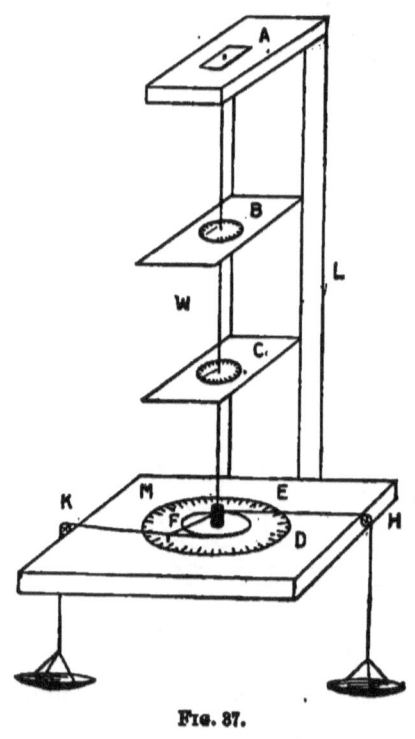

Fig. 37.

THE PROPERTIES OF MATTER. 55

IX. CAPILLARY ACTION.

77. Apparatus. — Sewing-Needles, small Wooden Balls, Iron Wire, Soap Solution, Glass Rod, Glass Plates, Capillary Tubes, Tumbler, Earthen Plate, Balance, Linear Scale, etc.

78. Exercise. — Place a sewing-needle on the surface of a vessel of water. If carefully done, it will float. A hairpin bent up slightly at the points may be used to advantage in letting down the needle so that its two ends touch the water about simultaneously. Observe carefully the shape of the surface of the water about the needle. Estimate as well as you can the area of the cross-section of the depression as compared with that of the needle. A body floats on water when it displaces a volume whose weight equals that of the body. Does the needle do it?

Place two needles on the water in positions parallel to each other, and separated by a few millimetres. Let a drop of alcohol or ether fall on the water between them, and note the effect.

If the surface of the water is covered with a powder called lycopodium, quite large wires can be floated.

79. Exercise. — Float two wooden balls, of about 15 mm. diameter, near each other on water. Their surfaces should be freed from all oily matter by washing them with a solution of caustic potash. Observe the shape of the surface of the water, and notice how the balls act toward each other. When quite close together, let fall a drop of alcohol on the water between them.

Coat one of the balls with a thin film of lard, and repeat the experiment. Study the shape of the water surface.

Repeat after coating both balls with lard.

80. Exercise. — Construct of No. 24 iron wire a ring 6 cm. in diameter, a tetrahedron 4 cm. on each edge, and a cube 5 cm. on each edge, each wire form having a handle attached (Fig. 38). In 400 ccm. of cold water that has been

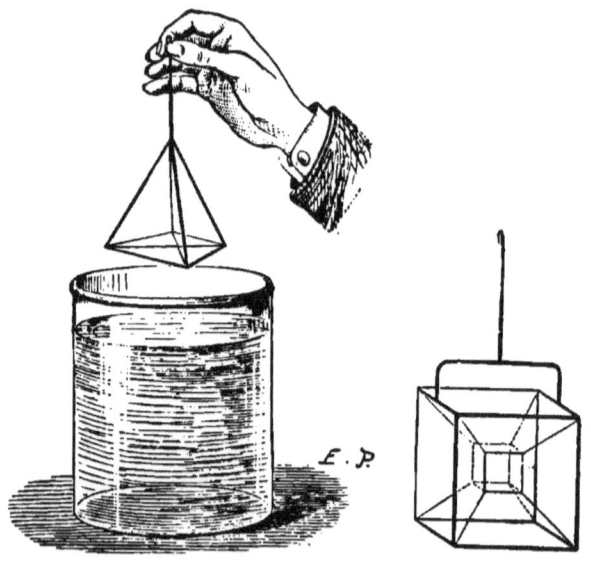

Fig. 38.

previously boiled, put 10 grammes of Castile soap cut up fine, or, better still, sodium oleate. Put this into a bottle, and set it in hot water over a gas flame turned low so as to keep the temperature about constant. Let it remain there an hour, shaking it occasionally till the soap is dissolved. Set aside for several hours in a place where it will not be disturbed, and then pour off the clear liquid, and add to it 270 ccm. of the best glycerine, shaking the whole thoroughly. Suspend in the wire

THE PROPERTIES OF MATTER. 57

ring a small loop of fine silk thread, and dip it into this soap solution. Break the film in the silk loop by puncturing it with a hot wire, or a piece of blotting-paper, and account for the shape it takes. Dip the other wire forms in this solution, and study the film forms obtained. Blow a soap-bubble with a common clay pipe, detach the bubble, and support it on the wire ring. Bring in contact with the bubble a second ring, and you will be able to draw out the bubble into a cylinder.

81. Exercise. — It is required to measure the tension of a soap film.

Support horizontally a knitting-needle or stout wire, AB (Fig. 39). Cut from a straight, slender straw a piece of uniform size and 10 cm. long, as CD. Attach to one side and at the centre, a light scale-pan made of paper. Find the weight of the straw and the attached pan.

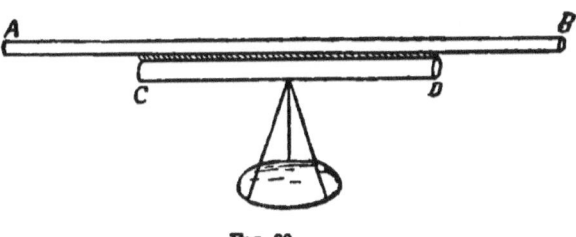

Fig. 39.

Hold the straw against the under side of the knitting-needle, and with a small brush introduce a layer of the soap solution between them. The tension of the film will now support the straw. Carefully sift fine sand on the pan till the film breaks. The weight of the sand, straw, and pan measures approximately the tension of the film. This divided by twice CD will give the superficial tension of either surface per unit of length.

82. Exercise. — Study the effect of a solution of camphor on surface tension by placing a piece of camphor of the size of the head of a pin on clean cold water, and noticing its peculiar movements. Touch the water with the finger, slightly oiled by rubbing it on the hair, and mark the effect. Using clean water, study the movements of a drop of a solution of camphor in sulphuric acid. Add other drops, and observe their behavior towards one another. Using clean water, study the action of a drop of a solution of camphor in benzole. Repeat all of these experiments, using warm water in place of cold.

The effect of camphor films on the surface tension can be learned by covering the surface of the water with lycopodium powder, then lowering a fragment of camphor into the water, and observing the effect that the camphor's touching the water has on the powder.

83. Exercise. — Suspend, by a thread, from a suitable support, a glass rod, so that the lower end dips into a tumbler of water colored with aniline dye or ink, the rod having a vertical position. Study the form of the surface of the water around the rod.

Repeat the experiment, using a glass rod thinly coated with oil.

Repeat these experiments, substituting mercury for water.

Thoroughly clean a strip of zinc by scouring it with a cloth after it has stood a few minutes in dilute sulphuric acid. Suspend it vertically in a dish of mercury.

Diagram the appearance of the surface of the liquid about the rod in each of these experiments.

Lift the rod slowly out of the liquid, observe the liquid just as the end of the rod separates from it, and then see if you can account for the various results obtained in these experiments.

84. Exercise. — Cut two plates of glass about 10 cm. square, and thoroughly clean them. In a large dinner-plate,

THE PROPERTIES OF MATTER. 59

pour some water highly colored with ink or aniline dye. Support one of the plates in a vertical position in the liquid. Examine the surface of the liquid about the plate. Now place in the liquid the second plate, securing it in a position parallel to the first, and separated from it by about 5 mm. Note the position of the surface of the liquid between the plates. Determine the effect of decreasing the distance between the plates.

Vary the experiment by making the plates touch along two vertical edges, the opposite ones being separated by about 2 mm. This is readily done by fitting a piece of wood between the two edges to be kept apart, and then slipping an elastic band around the two plates.

85. Exercise. — Determine the laws of capillary action.

Cut three capillary tubes of different diameters, 10 cm. long. Measure their diameters as in Art. 34. Construct a paper scale graduated to half-millimetres, cutting one end to a V shape, the zero of the scale being the point of the V. Coat the scale with transparent varnish, or with a thin film of paraffine. Fasten a tube to the scale by means of rubber bands, and then support the apparatus in a vertical position so that the tube dips into water, the zero of the scale just touching the water. Now raise the vessel containing the liquid about 1 cm., and then lower it to its original position in order to wet the interior of the tube with the liquid. Using a pocket lens to insure greater accuracy, read off the height of the liquid in the tube. This reading should be taken to the bottom of the meniscus, and increased by one-sixth of the diameter of the tube. When the heights have been ascertained for the different tubes, multiply each result by the diameter of the corresponding tube. If the measurements

PRACTICAL PHYSICS.

have all been carefully made, the products will be found to be nearly equal. What law of capillarity is here shown?

Using one of the tubes, determine the heights to which the following liquids rise: alcohol, benzine, turpentine, etc. The tube must be thoroughly cleaned after each experiment (see Art. 34), and the inner surface must also be wet with the liquid employed, as was done in the case of water.

A strip of plain sheet-metal may be used instead of the graduated paper scale. The position of the lowest point of the meniscus might then be carefully marked on it with a sharp-pointed instrument, and the distance afterwards measured with the dividers and scale.

Enter the results in the note-book as follows: —

CAPILLARY MEASUREMENTS.

Problem. — Determination of laws governing capillary action. Sept. 20, 1888.

Method. — [To the pupil. — Give brief summary of the process used.]

Results. —

	Length of Mercury Filament.	Weight of Vessel.	Weight of Vessel and Mercury.	Weight of Mercury.	Area of Cross-Section.	Diameter of Tube.	Height Water rises.	Height corrected for Meniscus.	Product of Diameter by Height.
First tube									
Second tube . . .									
Third tube									

X. SOLUBILITY.

86. Apparatus. — Small Flasks, Test-Tubes, Thermometer, and Balance.

87. Exercise. — Determine the solubility of a substance, that is, the number of grammes which one gramme of the solvent will dissolve, the solution to be saturated.

Fill a small flask, or a large test-tube, half full of the solvent. Add, in a powdered state, a quantity of the substance to be dissolved, being sure to add more than will dissolve. Now raise the temperature somewhat higher than that at which the solubility is to be determined; if water, it may be brought to the boiling-point. When the undissolved part no longer diminishes in quantity, remove from over the source of heat, and cool it to the required temperature by placing the vessel in melting ice. When the required temperature is reached, drop in the solution a crystal of the substance, and there will be precipitated all the salt in solution over and above that held in solution at the saturation point. Pour off into a small Florence flask of known weight some of this solution, weigh it accurately, and then, by placing it on a sand-bath, over a lamp or steam-coil, evaporate all the water, even that of crystallization, being careful not to hasten too rapidly the evaporation for fear of loss of material through "spitting." When thoroughly dry, weigh the residue, and obtain the weight of the substance dissolved. Divide the weight, in grammes, of the amount of the substance dissolved, by the amount of the solvent evaporated out of the solution, and the solubility of the substance is obtained.

88. Exercise. — Find the solubility of alum in water at 0° C., 20° C., 40° C., etc., to 100° C. Construct the curve of solubility. See Art. 598.

89. Exercise. — Find the solubility of potassium chloride in water at 0° C., 20° C., 40° C., etc., to 100° C. Construct the curve of solubility.

90. Exercise. — Determine the solubility of potassium nitrate in water at 0° C., 20° C., 40° C., etc., to 100° C. Construct the curve of solubility.

91. Exercise. — Determine the solubility of common salt in water at 0° C., 20° C., 40° C., etc., to 100° C. Construct the curve of solubility.

92. Exercise. — Determine the solubility of alum, potassium chloride, potassium nitrate, and common salt, in alcohol (95 per cent), at 20° C.

93. Exercise. — Determine the solubility of potassium sulphate in water at the temperature of the room; also of sodium nitrate. Finally, determine the solubility of sodium nitrate in the saturated solution of potassium sulphate.

THE PROPERTIES OF MATTER. 63

XI. DIFFUSION.

94. Apparatus. — Test-Tubes, Thistle-Tubes, Battery-Jar, Four-Ounce Bottle, Parchment Paper, Dialyzer, Porous Cup, Large Glass Tube, Sheet Rubber, etc.

95. Exercise. — Fill a large test-tube two-thirds full of water colored with blue litmus. Place a thistle-tube in the test-tube (Fig. 40), having it reach to the bottom, and pour into it a few drops of sulphuric acid. A reddish color will be seen at the bottom of the test-tube. Set the test-tube to one side, and record, from time to time, the position of the upper surface of the red-colored liquid.

Into a common test-tube, pour a little blue litmus solution, and add to it a drop of sulphuric acid by stirring, observing the effect on the color. Apply the truth taught by this to the explanation of the first. Account for the surface which separates the colored liquids not being stationary.

FIG. 40.

96. Exercise. — Measure the rate of diffusion of a substance in water.

Obtain a cylindrical vessel of about 2 litres capacity, — a common battery-jar will answer; a wide-mouthed bottle of about 4 ounces capacity, to be known as the diffusion-bottle; and a solution of the salt under investigation of some known strength, say, 6 parts by weight in 100 parts of water.

Fill the diffusion-bottle with the solution to be tested up to the brim, and place it on the bottom of the large cylindrical jar. Then pour water into this jar till it rises about 3 mm. above the top of the diffusion-bottle (Fig. 41). Let the

Fig. 41.

apparatus remain undisturbed for twenty-four hours, and at a constant temperature. Then, by means of a siphon, run off the contents of the jar into a basin, in which wash off the outside of the diffusion-bottle by means of a little jet of pure water, first taking the precaution to place a small plate of glass over the mouth so as not to spill any of its contents into the basin. Evaporate to dryness the solution in the basin, and determine the weight of the residue. Comparing this with the result obtained by using a solution of some other salt of the same strength will give their relative diffusion.

97. Exercise. — Compare the rates of diffusion of potassium chloride and potassium sulphate.

98. Exercise. — Ascertain if the strength of the solution in any way affects the rate of diffusion by using four solutions of common salt made by dissolving 1, 2, 3, and 4 parts of salt in 100 parts of water by weight.

These four solutions may be tested simultaneously by selecting jars and bottles of the same size. The diffusion-bottles especially must have the same capacity, and the openings must be of the same size.

99. Exercise. — Wet a piece of parchment paper, and tie it across the top of a large thistle-tube. Pour down the stem of the thistle-tube a concentrated solution of copper sulphate, spilling none on the outside of it. Now support the tube by means of a burette-holder in a beaker of water, so as to have the liquids at the same level on both sides of the tube to prevent hydrostatic pressure. Set aside for a few hours, occasionally observing the relative levels of the two liquids. Explain.

100. Exercise. — Set up an apparatus similar to that of the last experiment, and substitute a thin starch paste for the salt solution. Ascertain whether any of the starch passes through the parchment by testing the water on the outside with iodine water. A blue color would indicate that there was starch present. Iodine water is made by dissolving iodine in water.

101. Exercise. — Separate a mixed solution of common salt and starch.

Make a wooden hoop about 8 cm. in diameter and 5 cm. deep. Stretch across it a piece of parchment paper, making a shallow tray. Such an apparatus is known as a **Dialyzer**. Pour the given mixture into this tray, and float it on pure water in a large battery-jar or suitable vessel. Set aside for a few days, and then test the liquid in the jar for starch, as in Art. 100, and also for salt, by adding a little of it to a dilute solution of silver nitrate. A flocculent white precipitate, changing, when exposed to light, to a dirty pink color, would indicate the presence of salt. Pure distilled or rain water must be used.

FIG. 42.

A good form can be made by cutting off a large bottle (see Appendix, p. 354) about 5 cm. below the neck (Fig. 42).

102. Exercise. — Cement a small porous battery-cup to a funnel-tube with sealing-wax, making it air-tight at the line of junction. Attach the diffusion-tube to a two-necked flask which has a jet-tube extending through the second neck, and reaching below the surface of some water contained in the flask (Fig. 43). Fill a large bell-jar with hydrogen, and invert it over the porous cup. Account for what takes place. Now remove the jar from over the porous cup, and note the effect. Explain.

For method of preparing hydrogen, consult " Shepard's Chemistry," Exp. 24.

Fig. 43.

103. Exercise. — Determine the rate of diffusion of a gas.

Select a glass tube about 20 cm. long and 25 mm. internal diameter. By means of emery-powder and water on an old stove-lid or piece of flag-stone, grind one end so that a glass plate will close it gas-tight. Mix plaster-of-Paris with water to a thin paste, and out of it, on a glass plate, cast a disk about 3 mm. thick. When thoroughly dry, cut it with a sharp knife to fit the ground end of the tube, cementing it in with sealing-wax applied to its edges. The outer surface of the plaster plug should be about 2 mm. below the edge of the tube. Avoid getting any wax on the face of the plug. Paste a millimetre scale lengthwise of the tube, having its zero at the inner surface of the plaster plug, and give it a coat of thin varnish to protect it from water. It would be preferable to etch a scale on the glass. (See Art. 14.) Over the closed end of the tube, place a glass plate coated with a film of lard to close it gas-tight; then, holding the open end of the tube in the water of the pneumatic trough with one arm of a U-shaped glass tube passing up within the tube nearly to the plug, lower the cylinder slowly into the water. The air will escape through the U-tube, and the cylinder will be nearly full of water. If the plaster disk should get wet, it will have to be dried before proceeding further with the experiment. Remove the U-tube by dropping it down into the water-tank. Now take the reading of the water-level in the tube to obtain the amount of air left in it. Then, with a rubber tube, introduce the gas to be examined, filling the tube down to the level of the water in the tank. Take the reading again; the difference of the two will be a measure of the amount of gas. Now remove the glass cap, and, as the water rises in the tube, lower the tube to keep the level the same without as within to avoid hydrostatic pressure. When the water ceases to rise, take the reading of the water-level.

Let us suppose that the first reading was 2, and the second 100; then 98 measures the amount of gas introduced, and 2 measures the amount of air left in. If the last reading taken was 25, then 25 − 2, or 23, measures the air which passed through the porous plug, as against 98, which measures the amount of gas which passed out in the same time. Hence $\frac{98}{23} = 4.26$ is the ratio of the diffusion of the gas under examination to that of air.

104. Exercise. — Determine the rate of diffusion of hydrogen; also of oxygen.

For method of preparing oxygen, consult "Shepard's Chemistry," Exp. 7.

105. Exercise. — Tie a rubber membrane over the mouth of a glass vessel filled with oxygen gas, and place it under a bell-jar filled with hydrogen gas, standing on a glass plate. Account for what takes place. Try air and common illuminating gas.

Fig. 44.

106. Exercise. — Fill two wide-mouthed bottles with hydrogen and oxygen respectively, having previously fitted to them perforated corks through which passes a glass tube about 60 cm. long. After connecting these two bottles by the glass tube, set them in a vertical position (Fig. 44), with the one containing the hydrogen uppermost. After half an hour, remove the corks, and apply a lighted taper. If a loud explosion follows, it indicates that the gases have mixed. Why place the hydrogen-jar uppermost?

CHAPTER II.

MECHANICS OF SOLIDS.

I. LAWS OF MOTION.

107. Apparatus. — Wooden, Lead, and Glass Balls, Air-Pump, Spring-Balances, Electro-Magnet, and certain special appliances to be made in accordance with directions given hereafter.

108. Exercise. — Procure two balls of light wood of about 25 mm. diameter, place one of them near one end of a strip of common hemp matting stretched on the floor, and suspend the other by a string from some suitable support so placed that the ball just touches the one resting on the matting at a point a little above the extremity of a horizontal diameter; that is, the suspended ball must swing clear of the floor. Now pull to one side the suspended ball till it is, say, 15 cm. from the floor, and let it fall toward the other ball, striking it, and setting it in motion. Measure the distance the ball rolls on the matting before stopping. Obtain the average of several trials.

Repeat the experiment, substituting a piece of carpet for the matting. The suspended ball must be raised the same distance as before, in order that the force of the blow given to the ball may be the same as before.

Repeat the experiment on the floor; also try as smooth a surface as may be available. Tabulate all the results.

On comparing the distances the ball moves in the different cases, under equal impulses, with the character of the surfaces on which it rolls, to what conclusion do you come? What would you infer would be the result if a surface offering no resistance were used?

109. Exercise. — Find the resultant of two forces acting at an angle on a body at a point.

Procure two good spring-balances graduated to quarter-pounds, and a weight of about 10 pounds. Insert screws, at

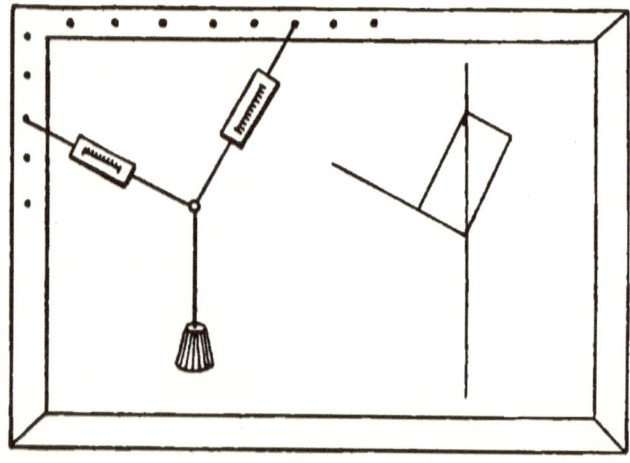

FIG. 45.

intervals of 3 or 4 inches in the frame, about one of the upper corners of the blackboard, for the attachment of wires to which the balances are to be fastened. The hooks of the balances are fastened to a small ring to which is also attached the known weight (Fig. 45). Trace lines of direction along the wires with a pencil. If accurately drawn, these will pass through the centre of the ring on being produced. After reading the indications of the balances, remove them, extend the lines, and lay off on them as many units of length as there are units of weight indicated by the balances respectively. Through the points thus located, draw parallels, making a parallelogram of which the line of direction of the weight will be the diagonal. On measuring this diagonal, it will be found to be 10 units long, thus showing, that, if we represent by lines the directions and the intensities of two forces acting at an angle, the diagonal of the parallelogram of which these lines are adjacent sides represents both the direction and the intensity of the resultant.

Attach the balances at different points in the frame of the board, and test the law stated above. Ascertain the effect of making the angle between the directions of the supporting wires very small. What do the balances indicate when both are attached to the same point, that is, the supporting wires are parallel? What does this show to be the value of the resultant of parallel forces? Increase the angle between the lines of direction of the two supporting wires, and determine the effect on the readings of the balances. Make the angle as large as possible. What would you infer from the last experiment would be the effect if the supporting wires could be made to act in exactly opposite directions?

110. Exercise. — Procure three balls of wood of about 25 mm. diameter each, and perforated with a small hole. Attach each to a cord about 60 cm. long, and then suspend the balls from a piece of board supported in a clamp by inserting the cords through holes drilled in the board, the distances between them being such that each suspended ball just touches the other two. Now draw off two of the balls to equal distances, and let them fall at the same instant against the ball at rest. Observe the direction of its motion. For one of the balls, substitute an iron one of the same size, and repeat the experiment. Compare the direction in which the third ball moves, with that obtained in the first case. Ascertain the effect of changing the angle between the lines traversed by the falling balls. What is the teaching of this experiment?

Glass balls may be substituted for the wooden ones, the cord being attached to the ball by tying it into a stout cloth loop cemented to the ball by means of a pitch cement. See p. 357.

111. Exercise. — Cut out of a board a circular piece 30 cm. in diameter, and divide it by radii into five-degree sections (Fig. 46), and mount it on a wooden support. Make four wooden blocks of the form shown in Fig. 47, each carrying a pulley about 3 cm. in diameter. With small iron clamps, attach the four pulleys at different places on the circular board. Knot together four flexible cords, and place them so as to draw over the pulleys. To the free end of these cords, attach weights, adjusting both the weights and the position of the pulleys so that the knot rests over the centre of the board. Tapping the apparatus with the finger will assist in overcoming friction. The weights may be cut out of sheet lead, and have the form shown in Fig. 48, or scale-pans can be used, equal lead balls serving as weights.

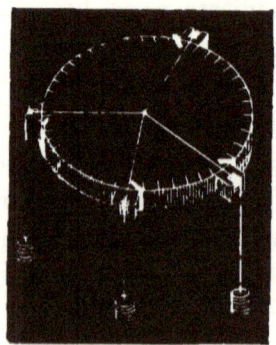

Fig. 46.

Fig. 47.

Now draw a circle on a piece of paper, lay off on it the position of the cords, and on each line measure off as many units of distance as there are unit-weights attached (Fig. 49). Find by the rules given for the composition of forces the resultant of any three of the forces represented, and compare this resultant with the fourth force as to direction and intensity. Inference.

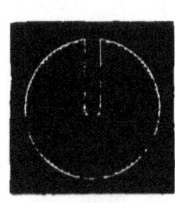

Fig. 48.

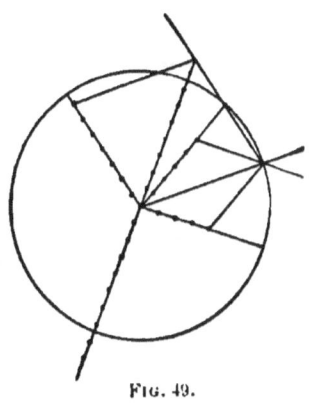

Fig. 49.

Vary the positions of the pulleys, and the ratios of the weights.

112. Exercise. — Determine the resultant of two or more parallel forces acting in the same direction.

Prepare a rectangular bar 27 cm. long, and having the uniform cross-section of 1 by 2 cm. At the middle, and at intervals 3 cm. each side of the middle, insert equal pieces of brass wire of about 2 mm. diameter and 3 cm. long (Fig. 50).

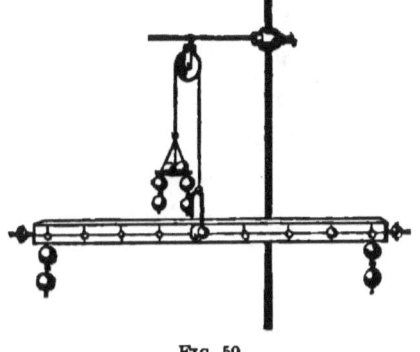

Fig. 50.

In each end, insert a piece of brass wire on which a small lead ball screws, to serve as adjusting weights to correct for inequalities in the wooden bar. Construct a small clevis out of brass wire to be used as a support to the bar. With common bullet-moulds, cast a quantity of lead balls. Drill small holes through them, and attach small wire hooks so that the balls may be attached to the bar. or to each other, as desired. A small pulley, a stout cord with a small ring or scale-pan attached to one end, and a suitable support, will be needed to complete the apparatus. Now, with the clevis, hang the bar by the middle pin, and counterpoise it over the pulley by means of weights. Move the adjusting weights till the bar is horizontal. Add weights at two points on opposite sides of the clevis, and place a sufficient number in the pan to balance the loaded bar. Compare the number added to the pan with the number attached to the bar. Observe the position of the clevis, as well as the positions of the pins to which the weights are attached, and then frame a law expressing these relations. Test the law by varying the positions and values of the weights.

A second device for determining the resultant can be constructed as follows : —

Procure a rectangular bar 1 metre by 5 cm. by 3 cm., of pine, graduate it to centimetres, screw on each end an ear made of sheet brass or iron, and suspend it by these ears in a horizontal position from the hooks of two spring-balances secured to some suitable support, as the edge of the laboratory table. Take the reading of each balance, so as to know what allowance to make for the weight of the bar. Now place on the bar a sliding-weight of known value, say 5 kg., and take the readings of the balances for different positions of the weight on the bar. Correct the readings for the weight of the bar. Examine the results for evidence of any relation between the corrected readings and the position of the weight. Compare the readings with the value of the weight.

MECHANICS OF SOLIDS.   75

113. Exercise. — Make two lead balls of about 3 cm. diameter. To do this, make a plaster-of-Paris mould, using a wooden ball as a pattern. Suspend them by cords from a horizontal bar so as just to touch (Fig. 51). Now draw one of them to one side, and let it fall against the other. How much matter moves after the impact, and how far does it move as compared with the distance the ball moved before the impact? Has the amount of motion been altered by the impact? What has been changed? What principles are suggested by this experiment? Raise both balls equal distances in opposite direction,

Fig. 51.

and let them collide, recording the effect. Is this result inconsistent with the laws of motion? Explain.

Bags of sand can be used as substitutes for the lead balls.

114. Exercise. — Suspend two elastic balls, ivory, wood, or glass, by cords, so as to swing in front of a graduated arc (Fig. 51) for convenience in comparing distances the balls move through. The balls, when at rest, should just touch each other. This arc may be cut out of cardboard, and equal distances measured off on it with a pair of compasses. Raise one of the balls through any number of spaces, and let it fall against the other. Observe the motion of the balls after impact. Harmonize the result with the third law of motion. Raise both balls through equal distances in opposite directions, and let them strike. Explain the result. Raise one through twice the distance of the other. Explain.

Repeat these experiments after fastening neatly around each ball a piece of flannel. Explain. Try lead balls. Try iron balls.

Are the balls perfectly elastic? What effects would be produced by perfectly elastic balls? What facts support this conclusion?

For a method of attaching cords to glass balls, see Art. 110. It is preferable to suspend each ball by two cords, the ball forming the point of a V, as the balls will then swing in a plane.

115. Exercise. — Suspend four or five highly elastic balls by cords, from a frame, in a manner similar to that described in Art. 114. Raise one of the balls through an observed distance on the scale, and let it fall against the adjacent one. Describe and account for the effect produced.

Substitute lead balls for the ones just used, or else cover them neatly with flannel or cotton-wool, and repeat the experiment. Why the difference? What would you infer would be the result if perfectly elastic balls were employed?

Consecutive balls must just touch each other when at rest.

116. Exercise. — Cut out of a board a semicircular piece having a radius of about 24 cm. At the centre of the semicircle, fasten with its face in the diameter a small rectangular piece of polished marble. Draw radial lines dividing the circle into ten-degree spaces. Suspend from a wire an ivory or other highly elastic ball, as glass, of about 2 cm. diameter, so as just to touch the marble surface exactly at the centre of the semicircular board (Fig. 52). Stand a common crayon near the outer end of any of the radial lines; draw off the ball, and release it over different radial lines till you find one by trial so situated, that, after the ball is reflected from the marble, it overturns the crayon. What law expresses the relation you find existing between the angles of incidence and reflection? Try a lead ball. Inference.

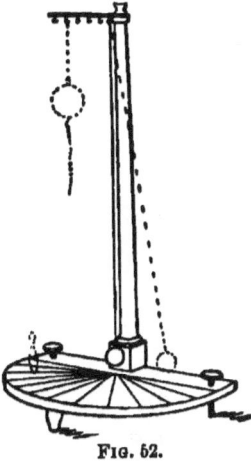

Fig. 52.

II. CENTRE OF MASS. — STABILITY.

117. Apparatus. — **Cardboard, Round-bottomed Flask, Double Cone, Cube, Prism, Pyramid, Pulley, etc.**

118. Exercise. — Find the centre of mass of a cardboard figure.

Cut a slot in a lead bullet with a knife; in the slot place a thread, and then close it up with a blow from a hammer. Cut from a sheet of cardboard a piece of any desired shape, and then suspend it from the edge of the table by inserting a fine needle through the cardboard as close as possible to the edge. Be sure that the figure is free to turn freely about the support. Now hang the bullet alongside of this cardboard figure, letting the string press gently against the needle (Fig. 53). Mark with a sharp-pointed pencil a point below the needle, and alongside of the thread. Through this point and the point of suspension, draw a straight line. In the same manner, locate a second line by suspending the cardboard from another point. Balance the cardboard figure on the point of a needle stuck into a cork for a base, and compare the position of the point when equilibrium is secured with the point of intersection of the two lines drawn. What position does the centre of mass sustain to the point of suspension when the cardboard hung from the needle in the edge of the table?

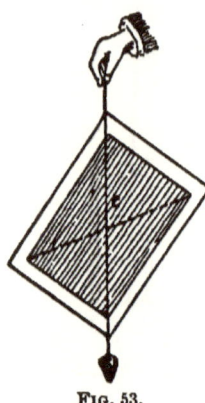

Fig. 53.

119. Exercise. — Find the centre of mass of the following cardboard figures: Square, rhombus, rectangle, circle, and triangle. What simple relation does this point sustain, in each case, to the form of the figure, enabling one to locate it by the aid of the ruler alone?

If the rectilinear figures are suspended by their corners, the law in each case will be readily seen.

120. Exercise. — Put a quantity of shot in a round-bottomed flask. Then fill the flask up with paper so that the shot will not move around on tipping the flask. Now overturn the flask, comparing its action on being released with that of one not so loaded. How is the centre of mass affected on overturning the flask? Where is the centre of mass when the flask is at rest?

MECHANICS OF SOLIDS.

121. Exercise. — Thrust a darning-needle, eye first, into a large cork, and try to make the apparatus maintain a vertical position, the needle-point resting on the point of a pencil. Account for the failure. Now thrust the blades of two heavy-handled knives into the cork (Fig. 54) on opposite sides, so as to form with each other an inverted V. Ascertain if the apparatus will now stand on the end of the pencil. Explain. Would it answer as well to have the knives perpendicular to the needle? Why? Whenever the apparatus stands alone, where is the centre of mass?

Fig. 54.

122. Exercise. — Turn out of wood a double cone 20 cm. long and 8 cm. in diameter. Make a V-shaped track (Fig. 55) out of two thin strips of board for the cone to roll on, giving it a length of about one metre, with the point 3 cm. lower than the wide end. Now place the double cone at the point of the track, and it will slowly roll toward the higher end. Does the cone roll up hill? Where is the centre of mass of the cone? Find, by measurement, whether the centre of mass has been raised or lowered by the motion.

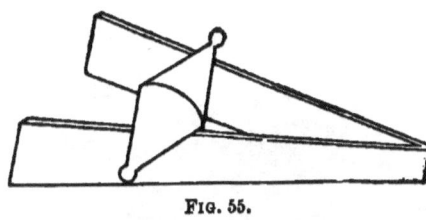

Fig. 55.

123. Exercise. — Determine the law governing the stability of bodies.

Cut out of hard wood, the heavier the better, a cube 1 dm. on each edge, a rectangular prism 1 by 1 by 2 dm., and a right pyramid whose base is 1 dm. square, and whose altitude is

2 dm. Insert a small screw-eye in a face of each solid opposite the centre of mass, and to it fasten a stout flexible cord. Stretch the cord over a pulley clamped to some adjustable vertical support, attach a scale-pan of known weight, and then add weights till the solid begins to overturn. In this way, compare the stability of these several solids as they rest on different faces successively, and generalize a law from the results. The height of the pulley should be such as to make the cord horizontal between the pulley and the solid; likewise, the base of the block must be secured against slipping.

III. CURVILINEAR MOTION.

124. Apparatus. — The Whirling-Machine and its Attachments.

125. Exercise. — Attach to a whirling-machine (Fig. 56) a strong frame, across which is stretched a wire carrying two equal balls free to slide on it, and connected by a thread.

FIG. 56.

Find, by trial, a position for these balls on the wire, such that, when the apparatus is rapidly rotated, the balls do not slide toward the end of the wire.

MECHANICS OF SOLIDS. 81

Repeat the experiment, using two unequal balls of known weights. How does the relation of the weights of the balls compare with that of their distances from the axis of rotation? What law of curvilinear motion is here illustrated?

126. Exercise. — Determine the laws for curvilinear motion.

Get a good mechanic to construct an attachment for the whirling-machine, of the form shown in the figure (Fig. 57), in which a ball is arranged so as to be clamped to a string which draws under a pulley, and thus lifts a weight whenever the ball moves away from the centre of rotation. The weights are lead disks of equal weight, provided with a slot, so that they can be placed on the platform attached to the cord. The weight of this platform may be taken as the unit weight. The centre of the weights must be exactly over the centre of rotation. Let us suppose that the ball is clamped at 20 cm. from the axis, and a load of 3 units is placed on the platform, making 4 units in all. Determine the number of revolutions the drive-wheel of the machine makes in, say, 10 seconds, in order just to

FIG. 57.

raise the load. As soon as the weight is seen to lift, then maintain a constant speed, and determine the rate. Use the average of several trials. Now replace the ball by one twice as heavy, and at the same distance from the axis; make the total load to be moved 8 of the unit-weights, and find the speed necessary to lift it. The number of revolutions will be found to be the same in each case. Repeat, using other distances and other weights, tabulating the results. Express as a law the relation between mass and centripetal force exhibited by these experiments.

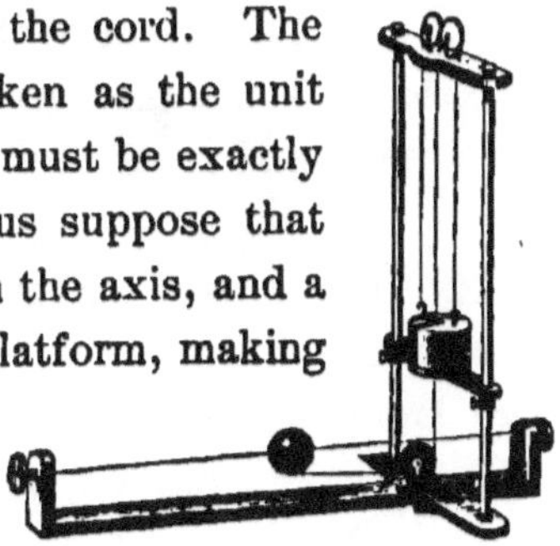

Again, set the small ball at 8 cm. from the axis, and determine, as before, the speed of the drive-wheel necessary to lift a load of 2 units. Compare it with that necessary to lift a load of 4 units when the ball is 16 cm. from the centre. Also compare it with that necessary to lift a load of 6 units when the ball is 24 cm. from the centre. How do the number of revolutions compare? Express as a law the relation shown to exist between the radius of the circle traversed by the ball, and the centripetal force or tension on the cord.

Again, set the small ball at 15 cm. from the axis, and determine the speed necessary to lift a load of 2 units. Compare this speed with that required to lift a load of 8 units. What measures the centripetal force in each case? Express as a law the relation existing between speed, or time of rotation, and centripetal force.

127. Exercise. — Attach to the whirling-machine a heavy ring having a diameter of about 30 cm. Suspend by a stout cord from the top of the ring, so as to hang in the axis of rotation, a small globe (Fig. 58), such as is used for aquariums, in which has been placed some mercury and some colored water. Rotate the apparatus, and observe the behavior of the liquids. Explain.

Fig. 58.

Suspend in a similar manner, in succession, the following: A skein of thread, a loop formed of a short chain, a prolate spheroid of wood, an oblate spheroid, a sphere, a double cone, and a cylinder.

Suspend these geometrical solids from the extremities of different axes, and ascertain if it makes any difference. Is there any law regulating their behavior?

IV. ACCELERATED MOTION. — GRAVITATION. — PROJECTILES.

128. Apparatus. — Smooth Plank, Atwood's Machine, Iron Balls, Electro-Magnets, Spring-Gun, etc.

129. Exercise. — Determine the laws of accelerated motion.

Take a straight plank about 5 metres long, and tack along the centre of it two strips, each having a cross-section of 1 cm. square, forming a narrow straight groove with sharp even edges not over 1 cm. wide. Graduate one of these strips in centimetres. Raise one end of the plank 40 cm. higher than the other (Fig. 59). Suspend a heavy ball by a thread 1 metre long, for a measurer of time. With a little adjusting, this pendulum will vibrate seconds. Now set the pendulum in vibration, and at the instant the ball reaches the lowest point of its arc, let an iron ball of about 3 cm. diameter roll down the groove on the plank, starting from the zero of the scale on the plank. Mark the point reached by the ball at the end of the first second. Make several determinations of this distance. In like manner, find the distance passed over in 2, 3, and 4 seconds respectively. What is the relation between the distance passed over in successive seconds? How is the total distance passed over at the end of any second related to the number of that second? How much is the acceleration?

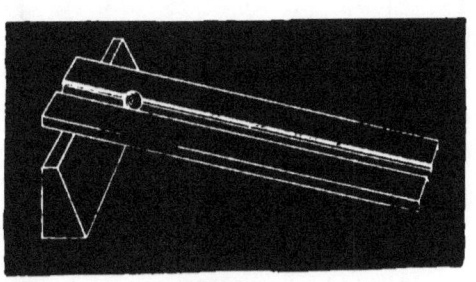

FIG. 59.

By what number would you multiply the acceleration in this case to give that for a body falling freely in space? Compare the result with the true value. Point out any source of error in the above method of determining the acceleration.

A stout wire stretched tightly across the room (Fig. 60) may be substituted for the plank. A simple way to stretch the wire is to fasten the upper end of it to the appliance commonly used in tightening the saw in the ordinary buck-saw frame. Place on this wire a pulley carrying a weight suspended from the under side. Parallel to the wire, and at a few centimetres above it, stretch a stout cord. Pieces of paper suspended from this so as to be moved by the pulley as it runs down the inclined wire will serve to measure distance.

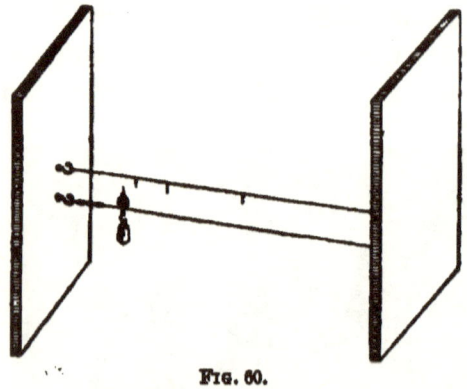

Fig. 60.

A metronome, such as musicians employ in measuring time, will be found very serviceable in this experiment.

As it is rather difficult to start the ball and pendulum simultaneously, it is recommended to hold the ball at the top of the inclined plane by means of an electro-magnet, with a piece of paper over the pole, the battery-circuit being closed by the iron bob of the pendulum being held in contact with the battery-poles when at the highest point of its path or swing. Then, on releasing the pendulum, the circuit will be broken, and hence the ball on the plane will be released at the same moment. The electro-magnet may be made by winding two or three layers of insulated copper wire

No. 16, on a piece of annealed iron 8 cm. long and 2 cm. in diameter.

The accuracy of the distance traversed by the ball can be tested by placing a ruler in the groove at the point, and then ascertaining if the click of the ball against the ruler is coincident with the expiration of the time as indicated by the pendulum.

130. Exercise. — Determine the laws of accelerated motion by means of an Atwood's machine.

A very efficient Atwood's machine may be made by a careful mechanic as follows: —

Mount an accurately balanced metal or wooden wheel of 20 cm. diameter on the top of a heavy wooden column (Fig. 61) set firmly into a triangular piece of plank, through which pass three wood or iron screws to serve as levelling-screws. A fine silk cord passes over the wheel, carrying two weights at its extremities. On the face of the column, graduate a centimetre scale 1.5 metre long, with the zero about 2 cm. below the lowest point of the wheel. Construct out of brass or iron a ring and a shelf, each provided with a clamp, so that they may be fastened to the column at any desired point in the path traversed by one of the weights. The weights can be made of lead or brass, and

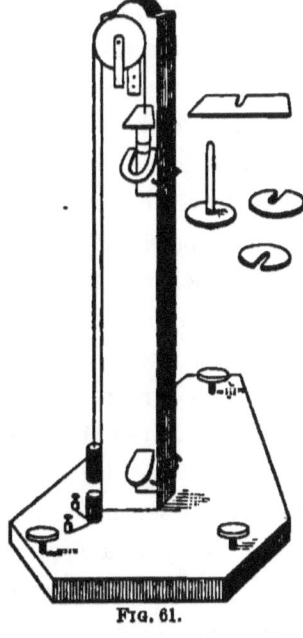
Fig. 61.

should be slotted disks of known weight. The platform of one of the weights must be made of iron in order that it may be attracted by the electro-magnet placed directly under it on the base. A metronome or clock beating half-seconds, so arranged that it can be put in circuit with a battery and the electro-magnet, will be found very desirable, as it will enable the experimenter to release the weight at the exact beginning of the interval of time marked by the clock. An iron ball suspended by a cord of a length to vibrate half-seconds can be used as a substitute, as in the last experiment. Slotted bars of lead or brass of known weight, called " riders " or " over-weights," to be removed by the ring-rest at any desired point, complete the apparatus.

First. Put equal loads on the two sides of the pulley. Set, by trial, the ring for arresting the over-weight so that the ring is reached by the descending weight in one second. Set the arresting-shelf so that it is reached at the end of the second second. The lower end of the descending weight starts from zero; hence its length must be added to the distance that the ring is below zero, to give the distance traversed in the first second. The distance between the ring and the shelf decreased by the length of the weight will be the velocity at the end of the first second. Why? Compare this distance with that passed over in the first second. Make several trials.

Secondly. Set the arresting-ring so that it is reached by the descending weight at the end of two seconds. Set the arresting-shelf so that it is reached at the end of the third second. What is the velocity at the end of the second second? Why? After determining, by repeated trials, the position of the arresting-ring and arresting-shelf, see if you can discover any fixed relation between the velocity at the end of any second of time and the number of the second.

MECHANICS OF SOLIDS.   87

Thirdly. Place weights on the pulley so as to differ by some known quantity; for example, use 275 grammes and 325 grammes, a difference of 50 grammes. Then the moving-force is due to this excess, and the mass moved is 275 + 325 + such a part of the mass of the pulley as, placed on the rim, would be equivalent to the whole mass distributed as now from the axis to the rim. In the case of a solid cylindrical pulley, this quantity is half of the total mass of the pulley. In the case of an open wheel, it is approximately the mass of the rim increased by the mass of the outer half of the spokes. This can be found by computing the volume in cubic centimetres, and multiplying it by the density of the substance.[1] Let the pulley be represented by 150 grammes. Then the total mass moved will be 275 + 325 + 150 = 750 grammes. Now determine, by trial, the position of the arresting-shelf so that it is reached in one second. Also determine the position of the arresting-shelf when it is reached in two seconds. Also find the position when reached in three seconds. What is the ratio of each of these distances to the first? If the total distance passed over is represented by s, and the time by t, what formula will represent the value of s in terms of t?

Fourthly. Find, as in the last case, the distance passed over by the descending weight in one second, two seconds, three

---

[1] The value of the pulley can be determined experimentally as follows: Let a = velocity generated in unit of time, w and w' = masses attached to the cord, w'' = mass of pulley when considered as placed at the circumference. Then $a = g \frac{w - w'}{w + w' + w''}$. Let s = space passed over by the descending weight in t seconds; then $s = \frac{w - w'}{w + w' + w''} \cdot \tfrac{1}{2}gt^2$. Place on the pulley the weights w and w'. Set w at the zero of the scale, and adjust by trial the arresting-ring to take off the overweight ($w - w'$) coincidently with any subsequent beat of the clock. The distance of the arresting-ring below zero will be the value of s, and the number of seconds elapsed from the beginning of motion till the ring is reached is the value of t. By substituting these values in the formula the value of w'' can be determined.

seconds, etc. Then, by taking their difference, we have the distance passed over during the *first* second, *second* second, *third* second, etc. Find the ratio that each of these distances sustains to the first.

131. Exercise. — Determine the increment of gravity, using Atwood's machine.

Let us suppose that a weight of 275 grammes is placed on each side of the pulley, and an overweight of 50 grammes on one side. If the pulley can be represented by 150 grammes, then the mass moved will equal 275 + 275 + 150 + 50 = 750 grammes, and the moving-force will be due to 50 grammes. The ratio of 50 to 750 equals the ratio of the moving-force to the force of gravity, and hence equals the ratio of the velocity the moving-force produces in a second to that gravity produces, that is, $\frac{50}{750} = \frac{k}{g}$, in which k is twice the space the 750 grammes move in one second. Now determine, by repeated trials, the position of the arresting-ring when reached in one second. Hence $g = k\frac{750}{50}$.

132. Exercise. — Find the measure of the effect of a force.

First. Using an Atwood's machine, put on one side of the pulley, say, 287.5 grammes, on the other 312.5 grammes. Hence, if the pulley is represented by 150 grammes, there is moved 750 grammes by a force due to the excess of 25 grammes of one weight over the other. This force acting on 25 grammes would produce a velocity of 9.8 m. in a second, and the momentum would be 25 × 9.8 = 245. This force moves 750 grammes, and as there must be an equality of momenta, then 25 × 9.8 = 750 × velocity acquired. Hence the velocity acquired by the 750 grammes will be $\frac{245}{750}$ = .326, and the space passed over during the first second will be

MECHANICS OF SOLIDS. 89

$\frac{.326}{2}$ = .163 metre; that is, if the arresting-shelf is placed at .163 metre below the bottom of the weight, it should be reached by the weight in one second. Test it.

Secondly. Using the weights 275 grammes and 325 grammes, giving a total mass of 750 grammes, and a moving-force double of the first, being due to 50 grammes, we have $50 \times 9.8 = 750 \times$ velocity acquired, velocity $= \frac{50 \times 9.8}{750} = .653$ metre, and space passed over in one second $= \frac{.653}{2} = .326$ metre. Therefore the arresting-shelf placed at .326 metre should be reached in one second by the descending weight. Test it.

It is now seen that a force due to 50 grammes imparts to 750 grammes twice the velocity that a force due to 25 grammes does; that is, while the mass remains constant, the velocity generated in a unit of time varies as the force.

Thirdly. Keeping the force which causes the motion the same as in the first case, by putting 100 grammes on one side, and 125 grammes on the other, making a total load of 375 grammes, one-half of the first load employed, we have $25 \times 9.8 = 375 \times$ velocity acquired, from which the velocity acquired $= \frac{25 \times 9.8}{375} = .653$ metre, and the space passed over $= .326$ metre. Test it.

Now the ratio of the masses moved in the first case and in the last is $\frac{750}{375} = 2$, and the ratio of the velocities produced is $\frac{.326}{.653} = \frac{1}{2}$. Hence, while the moving-force remains constant, the velocity generated in a unit of time varies inversely as the mass moved. Change the loads, and ascertain if this statement is general.

From these considerations, it is evident that the effect of a force may be measured by the product of the mass moved by the velocity produced, that is, may be measured by the momentum acquired in a second.

133. Exercise. — Ascertain if mass affects the velocity of falling bodies.

Procure two or three iron balls of different sizes, and as many U-shaped electro-magnets. Tie the electro-magnets to a stout stick so that they may be held out of an upper window of some high building. These electro-magnets should all be in circuit with a battery of sufficient strength to cause them to hold up the balls. By breaking the circuit, the balls will all be released at the same moment, and one standing on the ground can determine if there is any difference of time in their reaching the ground. Does a difference of weight perceptibly affect the time of falling?

Drop from your hand, at the same instant, an iron ball and a cork one, and ascertain if the conclusion in this case harmonizes with that previously reached. Then make a paper cone out of light cardboard, and of sufficient size to hold both balls. After placing the balls in the cone, the iron one first, drop it point first. Are the results the same as before? Explain.

Drop a coin and a paper disk of the same size simultaneously, comparing the result with that obtained by placing the paper disk on top of the coin, and dropping them from the hand. Now place the coin and the paper disk in a long glass tube sealed at one end, and having a stop-cock fitted to the other. Such an apparatus is supplied by dealers under the name of **Guinea and Feather Tube**. Connect the tube with the air-pump, and exhaust the air as perfectly as possible. Close off the tube, remove it from the pump, invert quickly, and ascertain if the relative time of falling of the two objects is affected in any way.

Prepare a set of propositions embodying the truths exhibited by these experiments.

134. Exercise. — Support a pail of water at an elevation of two or more metres above the floor. Place in the pail one end of a rubber tube of sufficient length to reach to the floor. In the other end of the tube, insert a short piece of glass tubing drawn out to a jet-point. See p. 356. Secure this tube in the jaws of a suitable clamp so that any desirable direction may be given to it. Start the water flowing through the tube by suction, and then observe the shape of the path of the stream, and the horizontal distance reached by it, as the direction of the jet-tube is changed. Inference. Ascertain the effect of a change in the height of the pail above the end of the jet-tube. What forces determine the path of the stream?

135. Exercise. — Determine the path of a projectile.

Construct a spring-gun after the pattern shown in Fig. 62. The horizontal and vertical pieces are 2.5 cm. thick; N is a cylinder with tenons of equal length sliding freely through round holes in the vertical pieces; K is a pin to serve as a handle; L and M are elastics or spiral springs firmly fastened to N and the vertical CD. Two wooden balls of the same diameter are needed. Through each, drill a hole of such a size that either will slide easily on the tenon F. The tenons on N should be of such a length, that, when N is drawn back, the right-hand one will be flush with the face of CD, and the left-hand one will project a distance equal to the diameter of the ball. Now draw N back, place one ball on F, and the other on the ledge CB opposite the hole at H; then release N, and it will be seen that both balls will be set in motion at the same time, one falling vertically, and the other moving as a projectile. AB must be held in a horizontal position. Compare the motions of these balls, and the times required to reach the floor.

To picture the path of the projectile, hold the spring-gun so that the path of the ball will be parallel to a vertical wall, and distant a few centimetres from it. By watching carefully the ball, it will be found possible to insert a pin in the wall opposite some position of the ball in its flight, new points being marked each trial, as well as those just located verified. A line traced through these points will represent closely the path of the projectile. This can be transferred to a sheet of paper by ascertaining, first, the vertical distance of each of these points from the horizontal line passing through the highest point of the curve, as well as the horizontal distance from the initial point of the curve; and secondly, locating them with reference to a corresponding line drawn on a sheet of paper, reducing each measurement on the same scale.

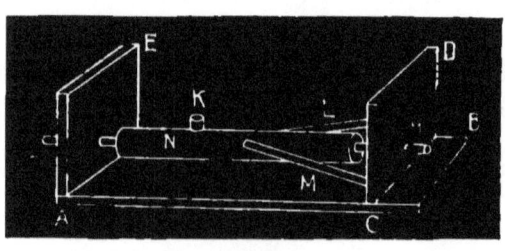

FIG. 62.

V. THE PENDULUM.

136. Apparatus. — A number of Balls to use in constructing pendulums.

137. Exercise. — Determine the laws for the vibrations of pendulums.

Fasten to each of a number of lead balls, of about 15 mm. diameter, a stout cotton thread, by cutting in each a slot with a knife, inserting one end of the thread, and closing it up by a blow from a hammer. Drill a number of holes, 5 cm. apart, through a piece of board about 60 cm. long, 5 cm. wide, and 15 mm. thick. Pass the strings through these holes, securing each in its place with a small wooden wedge. Support this board in a horizontal position in some suitable way (Fig. 63). These suspended balls will serve as pendulums, the lengths being the distances respectively from the under side of the support to the centre of the balls, and easily varied by pulling the strings through the holes in the support. Marbles may be substituted for the lead balls, the strings being attached by some very adhesive cement. See p. 357.

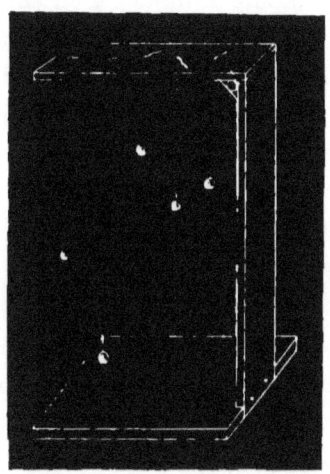

Fig. 63.

First. Set one of these pendulums swinging through a small arc, about five degrees, and determine the number of vibrations made in thirty seconds by averaging several trials. Now set the pendulum vibrating through a much larger arc, and determine the number of vibrations made in thirty seconds. Does the size of the arc affect the time? Test the conclusion by extending the time of observation, as well as increasing still more the size of the arc.

Secondly. Adjust, by trial, the length of one of the pendulums so that its time of vibration is one second. How long is the pendulum? Adjust another to vibrate in half-seconds;

another in one-third of a second. Compare the lengths of these several pendulums, ascertaining if there is any simple law connecting the lengths of the pendulums with their times of vibration.

Thirdly. Set up two pendulums of the same length, using a wooden ball for the bob of one, and a metal ball for the other. Compare carefully their times of vibration. Inference.

Fourthly. Using an iron ball as a bob for a pendulum, determine carefully its time of vibration. Now place just below the lowest point traversed by the ball, a pole of an electro-magnet, and again determine the time of vibration. Would you infer from the comparison of these times, that an increase in the force of gravity would affect the time of vibration of a pendulum?

138. Exercise. — Fasten six lead or iron balls on a stout string at intervals of 15 cm., forming a compound pendulum about 105 cm. long. Set the pendulum in vibration, and determine its time. Find the time of vibration of a pendulum whose length is the distance of the centre of the lowest ball of the six from the point of support. In like manner, find the times of pendulums whose lengths are the distances respectively of the other balls from the point of support. Set up a pendulum whose time is that of the compound pendulum. Notice the shape of the compound pendulum when vibrating, and apply the facts ascertained above to account for it. Which ball would you remove to shorten the time, and which to lengthen? Add another ball to the number without affecting the time. Explain.

139. Exercise. — Suspend from a suitable frame a uniform strip of wood about 1 metre long, 5 cm. wide, and 1 cm. thick,

MECHANICS OF SOLIDS.

by driving into the end of it a small wire staple, forming an eye through which a knitting-needle can be passed to serve as an axis of oscillation (Fig. 64). Adjust a pendulum, as in the last experiment, to vibrate in the same time as the wooden bar. Compare the length of the pendulum with that of the bar. Now cut from a sheet of lead a piece weighing half a kilogramme. Bend this into a clasp that will slide along the bar with sufficient friction to stay wherever placed.

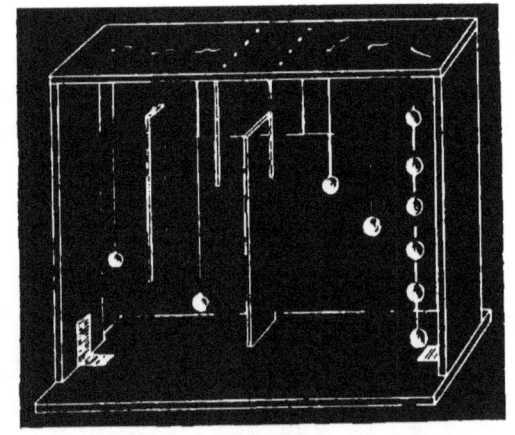

FIG. 64.

Having measured off on the bar the length of the pendulum vibrating in the same time, place the sliding-clasp below that point on the bar, and note the effect on its time of vibration. Note the effect when the weight is at the point, and also when above the point.

140. Exercise. — Suspend a wooden bar as in the last experiment, and adjust a pendulum to vibrate in the same time. Bore a hole through the wooden strip at the point marked in that experiment, and support the bar by passing the needle through it. How does its time of vibration compare with its previous time? State as a proposition the relation found to exist between this point and the centre of suspension.

141. Exercise. — Suspend a common lath by a cord 10 or 15 cm. long, so as to swing as a pendulum. Adjust a ball-pendulum to vibrate in the same time. Mark a point on the lath distant from the suspension-point the length of the ball-pendulum. Now set the lath in vibration by striking it with a ruler opposite this point. Compare the effect with that produced when the lath is put in vibration by a blow delivered at some other point on the bar. Inference.

VI. FRICTION.

142. Exercise. — Measure the friction between two surfaces when one moves over the other.

Procure a board about 1 metre long and 25 cm. wide, dressed to a plane, and a rectangular block 12 × 6 × 4 cm. In the centre of each of any three faces about a corner, insert a screw-eye for convenience in attaching a cord. Weigh the block, place on it any convenient number of grammes, attach a spring-balance by means of a cord; then, keeping the cord horizontal, apply sufficient force to move the block, reading the balance just as the block starts to move, and also after motion has begun. Make several determinations of both the starting and the moving force.

Repeat the tests to ascertain if the area of the face of the block resting on the board affects the results.

Ascertain the effect of using a greater weight.

Determine the effect of varying the character of the substances by placing a plate of glass, iron, brass, etc., on the board, and cementing to the face of the block a thin sheet or plate of the same substance.

Measure the friction between the surfaces when coated with some kind of lubricator.

MECHANICS OF SOLIDS.

Make several determinations in each case, and, using the mean as the most probable value of the force applied, divide it by the mass moved, to obtain the coefficient of friction.

As a substitute for the balance, the string may be stretched over a pulley by weights placed in a scale-pan.

VII. THE SIMPLE MACHINES.

143. Apparatus. — A Lever, several Pulleys, a Wheel and Axle, an Inclined Plane, and a Set of Weights.

144. Exercise. — Determine the law of equilibrium for the lever.

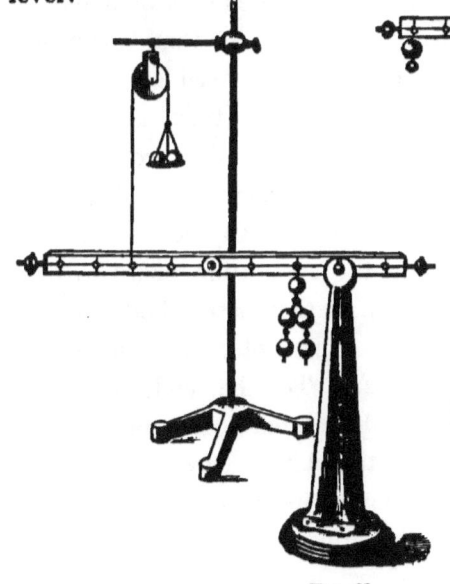

Fig. 65.

The apparatus employed in Art. 112 will be found useful for this one. When any other point than the centre of the bar is taken as the fulcrum, the weight of the bar can be neutralized by means of a lead ball made to screw on the end of the wire inserted in the end of the bar. A Y-shaped support (Fig. 65) will be found convenient, but it is not indispensable, as the wire clevis can be used to support the bar.

Set the fulcrum at any point, and counterpoise the bar. Suspend any number of weights from a point on one side of the fulcrum, and a sufficient number from a point on the opposite side to cause the bar to assume a horizontal position. Compare the ratio of these weights with that of the spaces between their points of attachment and the fulcrum. Vary the weights and their points of attachment. Arrange the apparatus so that one weight is between the fulcrum and point of application of the other weight. What law do you find expresses the relation between the weights and the arms of the lever?

145. Exercise. — Secure in a vertical position a board about 1 metre long and 40 cm. wide. Mount on this board a circular wooden disk of about 20 cm. diameter, so that it can turn freely about a fixed axis through the centre. Attach at any two points of the disk cords which pass over pulleys (Fig. 66) which can be given any desired position on the supporting board by having, as their axes, bolts moving in vertical slots, and fastened by nuts. To the ends of these cords, attach scale-pans cut out of tin plate. Now place known weights in these pans, and when the apparatus assumes a position of rest, measure the perpendicular distance from the centre of the disk to the line of direction of each cord, and compare their ratio with that of the weights. The weights should include the pans, and the position of the

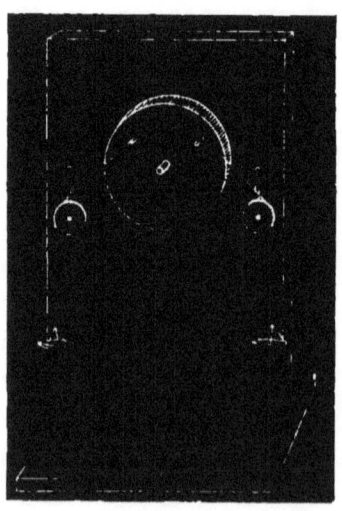

FIG. 66.

MECHANICS OF SOLIDS.

cords should be such that the direction does not pass through the centre of the disk. Make a number of experiments in which the weights and the positions of the pulleys are changed. Express as a law the relation found to exist.

146. Exercise. — Determine the law for the movable pulley.

With three small pulleys, some flexible cord, and two small scale-pans, set up successively the arrangements shown in Fig. 67. Place shot in one of the pans to counterbalance the weight of the pulleys, then place known weights in the two pans, and ascertain the ratio between them whenever equilibrium is secured. Account for the value of the ratio in each case.

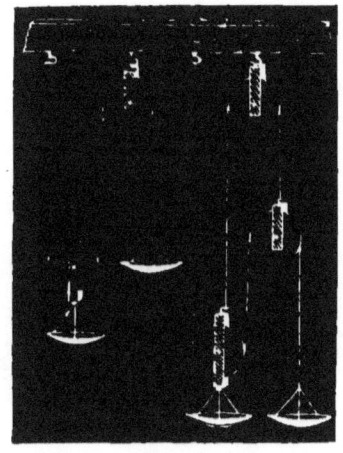

FIG. 67.

147. Exercise. — Determine the law of equilibrium for the wheel and axle.

A suitable apparatus for this experiment may be made by a good mechanic as follows: —

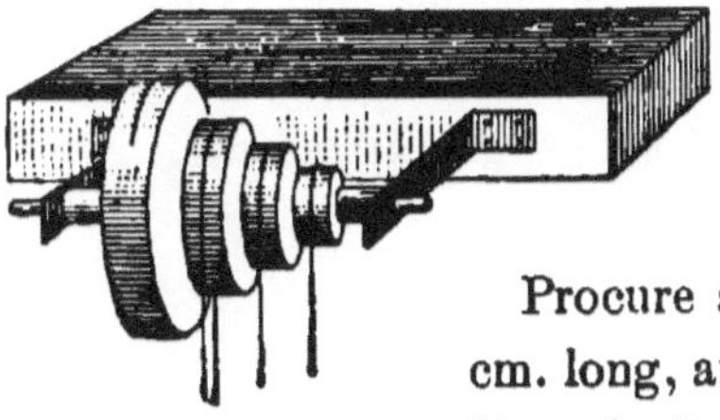

Procure a piece of hard wood about 15 cm. long, and of about the same diameter. Turn it down in a lathe so as to have a succession of disks with diameters as the numbers 1, 2, 3, etc. These disks need not be more than 2.5 cm. thick. Fasten to the circumference of each disk a cord for

FIG. 68.

the attachment of weights. The axis, consisting of a piece of brass or iron rod driven into the ends of the piece of timber, can be supported from two brackets firmly screwed to the edge of a piece of plank (Fig. 68). Attach scale-pans to the cords; place shot in one till equilibrium is secured, then, employing known weights in the pans, ascertain the law of the machine.

148. Exercise. — Determine the law for the inclined plane.

Construct an apparatus for this experiment as follows: —

Hinge together two pieces of board, each about 15 cm. wide and 50 cm. long. On the under side of the one to be used as the inclined plane, fasten thin strips across it at intervals of 2 cm. Then, by means of a prop moved along between the

FIG. 69.

boards, kept from slipping by the strips on the upper one, any desired elevation may be given to the plane. Fasten a pulley at the opposite end from the hinge, so that a cord attached to a car moving along the plane, and drawing over the pulley, will be parallel to the plane. Place weights in the scale-pan to counterpoise the car, then add weights to both the car and the scale-pan, comparing them, when equilibrium is secured, with the dimensions of the plane.

Fig. 69 illustrates a modified form of this apparatus.

CHAPTER III.

MECHANICS OF FLUIDS.

I. PRESSURE IN FLUIDS.

149. Apparatus. — Large Glass Tube, Two-litre Bottle, Metal Tube, Bladder-Glass, Weight-Lifter, Magdeburg Hemispheres, Air-Pump, Glass Tubing, Barometer-Tube, Aneroid Barometer, small Rubber Foot-Ball, Glass Bolt-Head, Model of Sprengel Pump, etc.

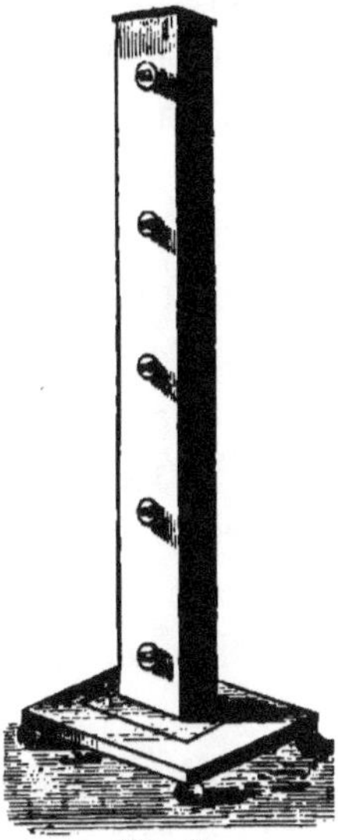

150. Exercise. — Close with a cork one end of a brass or tin tube about 1 metre long and 3 cm. in diameter, and set it into a wooden block for a base. In the side of the tube, drill five holes 5 mm. in diameter, at equal distances apart, and placed so that the third hole is at the middle of the tube (Fig. 70). Close these apertures with corks, and fill the tube with water. On uncorking the apertures, compare the ranges of the streams, and also ascertain if they differ in force. *Inference.* The water in the tube should be kept at a constant level while these comparisons are being made, by pouring water from a suitable vessel.

FIG. 70.

151. Exercise. — Ascertain if the atmosphere exerts pressure.

Tie firmly a piece of sheet-rubber over one end of a large tube, tin or glass, about 30 cm. long and 5 cm. in diameter. After filling it with water, invert it in a vessel of water, keeping the mouth just below the surface. Observe the shape of the rubber surface. Ascertain the cause of it by substituting a carefully fitted glass plate for the rubber, and comparing the force needed to pull it off when the tube is full of water with that when free from water.

152. Exercise. — Tie a piece of sheet-rubber or tough paper over one end of a bladder-glass, place it on the table of the air-pump, and exhaust the air. What fact is proved?

To make a bladder-glass, cut off the bottom of a common glass fruit-jar (see p. 354), and grind down the edges on a piece of flat sandstone.

Coat with lard the edges of the glass resting on the air-pump table.

153. Exercise. — Connect the **Weight-Lifter** (Fig. 71) to the air-pump by a piece of heavy rubber tubing. Account for the upward movement of the piston with the heavy weight attached as the air is exhausted from the cylinder. Measure the cylinder, and compute how great a weight can be lifted by the machine. Remove the piston, and tie a rubber membrane across the end of the cylinder. Observe the effect on the membrane on exhausting the air as different positions are given to the cylinder, varying from vertical to horizontal. Inference.

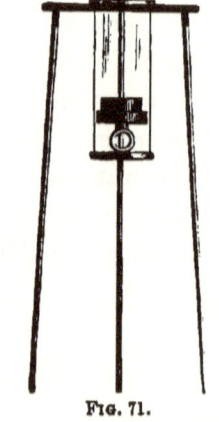

FIG. 71.

MECHANICS OF FLUIDS.   103

154. Exercise. — Connect the **Magdeburg Hemispheres** (Fig. 72) with the air-pump. Exhaust the air, remove them from the pump, and then try to separate them. While thus fastened together, place them under a bell-jar on the table of the air-pump, exhaust the air, and observe the effect. What do you find held the hemispheres together? Ascertain if the hemispheres can be pulled apart in any one position more easily than in another. Inference.

Fig. 72.

155. Exercise. — Compare the pressure exerted by fluids in different directions.

Bend a stout glass tube about 75 cm. long into the form shown in Fig. 73 (a), and also three shorter pieces into the forms shown in (b), (c), and (d). In the U-shaped part of (a), pour enough mercury to rise about 1 cm. in each arm. Attach the tube (b) to the lower end of (a) with a rubber connector, and hold the apparatus vertically in a vessel of water, observing the distance of the mouth of (b) below the surface of the water, and also the effect on the level of the mercury in (a). Now substitute the tube (c) for (b), submerging the apparatus to the same depth as before, as indicated by the mouth of the tube. Note the change in the mercury level. Try (d). What causes the change of

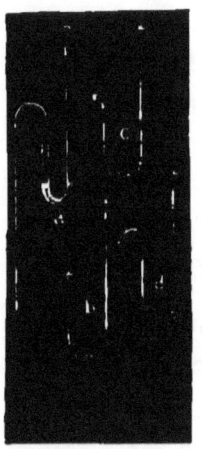
Fig. 73.

level? What is the evidence? What truth is proved by this experiment?

The tubes (b), (c), and (d) must be filled with water as high as the mouth, to insure success.

156. Exercise. — Cut a board about 75 cm. long, 15 cm. wide, and 15 mm. thick. Place it on a table so that it projects 25 cm. over the edge. Spread a large newspaper over the end on the table, making it as smooth as possible. Now strike the projecting part of the board with the hand, comparing the blow necessary to raise the other end of the board with that required when the paper is removed, or a smaller paper is used. Explain.

157. Exercise. — Select two test-tubes, such that one slides easily within the other. Fill the larger one with water, insert the closed end of the smaller tube, and quickly invert. Account for the upward movement of the small tube in opposition to the force of gravity. What is the office of the water in this experiment?

158. Exercise. — Measure the atmospheric pressure.

Fill a heavy glass tube about 80 cm. long, and closed at one end, with mercury. To do this, connect to the tube a small glass funnel by means of a rubber connector. Invert it in a cup of mercury (Fig. 74) by placing the finger firmly over the end of the tube to keep the mercury from running out during the operation. Why does not the mercury all run out of the tube? Devise some way of showing that it is atmospheric pressure on the mercury in the cup that supports the column in the tube. Measure the height of the column in the tube above the level of the mercury in the cup. Incline the

MECHANICS OF FLUIDS. 105

tube, and again determine the difference of level. Why the same?

Measure the cross-section of the tube, and then compute the volume of mercury in cubic centimetres. Multiply this volume by 13.6, and you have the weight of mercury in grammes. Compute what the weight would be if the cross-section of the tube were 1 scm., and the mercury column 76 cm. long.

In filling the tube with mercury, remove air-bubbles by sliding down the tube a slender iron wire.

159. Exercise. — Measure the height of a building or hill by means of a barometer.

For this purpose, use an **Aneroid Barometer** (Fig. 75), as it is more portable and more sensi‑

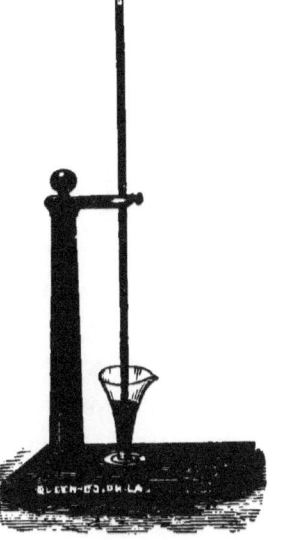

Fig. 74.

tive than the mercurial. Note the rise or fall of the barometer in hundredths of an inch in passing from one station to the other, multiply by 9, and the product is the difference in altitude expressed in feet. If the pressure is below 26 inches, or the temperature above 70° F., use 10 for a multiplier. Many readings at both stations should be taken, extending over a period of several days, the averages being used in the computation.

Fig. 75.

The readings should be taken with the aneroid in a horizontal position, tapping the face slightly with the finger before taking them in order to bring the needle into equilibrium. Guard against parallax in taking the readings, and estimate all fractions of spaces.

For more accurate methods of measuring heights, consult Plympton's "Aneroid Barometer," published by D. Van Nostrand, New York.

160. Exercise. — Fill a small rubber football half full of air, and place it under a bell-jar on the air-pump table. Exhaust the air from the jar; notice the effect on the ball. Explain.

Substitute a dish of soap-bubbles for the ball, or an empty bottle with its mouth opening under water in a tumbler.

What property of air do these experiments show?

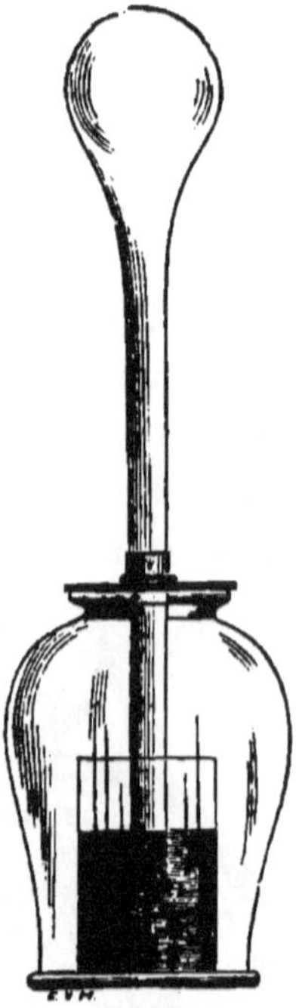

161. Exercise. — Procure a bolt-head, that is, a stout glass tube with a large bulb on one end. A strong glass tube fitted air-tight to a bottle by means of a perforated cork will make a good substitute. Close the top of an open-top bell-jar with a good cork, through which passes this glass tube reaching to the

FIG. 76.

bottom of a vessel of water within it (Fig. 76). Place the apparatus on the table of an air-pump, exhaust the air, and watch the effect. Also observe the effect of admitting the air. Explain.

MECHANICS OF FLUIDS. 107

162. Exercise. — Fit to a flask or bottle a good cork, through which passes a jet-tube. Connect this tube to an air-pump by means of a rubber tube, and exhaust the air. Close the connection between the flask and the pump with a pinch-cock, disconnect the tube from the pump, and open the pinch-cock after placing the mouth of the rubber tube in a vessel of water. Explain.

163. Exercise. — Connect with rubber tubing, as shown in Fig. 77, a glass funnel, a T-tube, a stout glass tube one metre long, and a flask to be exhausted. On the rubber connector, between the T-tube and the funnel, place an adjustable pinch-cock. Secure the apparatus in a vertical position, with the lower end of the tube fitting into a vessel which has an opening on the side a little higher than the end of the tube. Now close the pinch-cock, pour mercury into the funnel, and open the pinch-cock on the tube leading to the vessel to be exhausted. Then open the pinch-cock so that the mercury falls in a rapid series of drops, closing the tube, and preventing air from entering from below. As the mercury flows down, the exhaustion begins, the air of the tube being carried down by the mercury drops, and the air of the flask expanding to fill its place. The mercury can be poured back into the funnel from time to time, care being exercised to prevent the funnel's

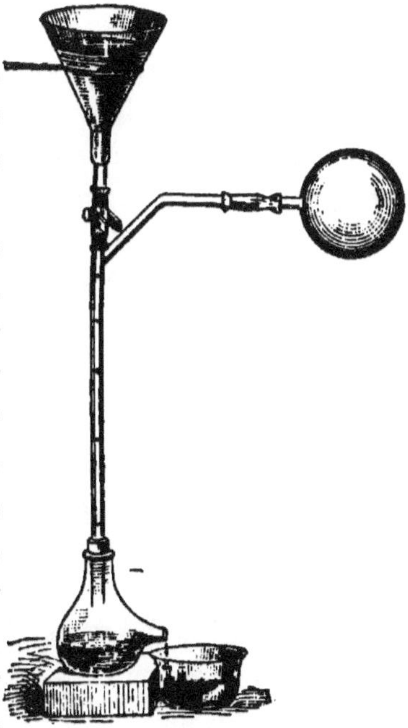

FIG. 77.

becoming empty, as then air would immediately enter the flask. Search for evidence in the working of this mercury-pump, showing that the degree of vacuum is improving.

II. LAW OF BOYLE.

164. Apparatus. — Heavy Glass Tubing, Mercury, etc.

165. Exercise. — Determine the law of the compressibility of gases.

First. Close one end of a stout glass tube in a gas-flame, and then bend it into the shape of the letter J, the short arm being the closed one, and about 25 cm. long (Fig. 78). Lengthen the long arm to 2 metres, using a heavy rubber connector, and winding it with fine copper wire. Fasten the completed tube with metal straps to a board set into a block for a base. Attach metric scales to the arms, placing the zero of each one on the same level, and just above the bend in the tube. Now pour in enough mercury just to close the bend up to the zero of the scales, bringing the mercury to the same level by jarring the tube. Under what pressure is the air in the short arm? After this adjusting has been carefully done, pour in mercury till a reading of about 20 cm. is secured in the long arm. Record the height of the mercury column in each arm. Add more mercury and record the readings, proceeding in this way till the long arm is full of mercury. The difference of any two corresponding readings, plus the barometer reading at the time of conducting the experiment, will measure the pressure on the confined air-column. Why? Compute the pressure for each observation and multiply it by the length of the confined air-column corresponding to it. If the experiment is carefully conducted these products will be found to be nearly equal.

Tabulate the results.

Express as a law the relation between volume and pressure exhibited by the experiment. Represent this relation by a curve (598).

The mercury must be poured in the tube very gently, and with the tube in an inclined position at first, so that the mercury stream will not act as a piston, and drive air ahead of it into the short arm.

Secondly. Fill with mercury to within 20 cm. of the end a stout glass tube 80 cm. long, closed at one end. Under what pressure is the 20 cm. of air in the tube? Close the end of the tube with the finger, and invert it in a vessel of mercury. Accurately measure the length of the air-column, and also the height of the mercury-column, above the level of the mercury in the vessel. What supports the mercury in the tube? Observe what the barometer-reading is. Why is the mercury-column less in height than that in the barometer? Under what pressure must the air in the top of the tube be? How does the ratio of the initial and present volume of the air in the tube compare with that of the pressures it is under in the two cases? Make several experiments with different initial volumes of air, and ascertain if the same relation holds good. Express as a law the relation discovered.

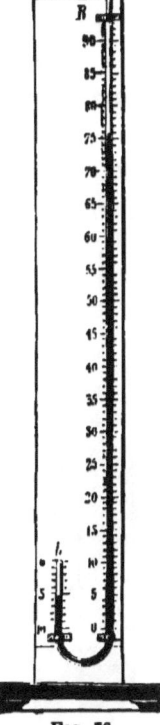

FIG. 78.

In order to subject the same volume of air to different pressures less than one atmosphere, a deep cistern of mercury will be required. To provide this, close one end of a heavy glass boiler-tube about 75 cm. long and 2.5 cm. diameter, and insert it in a heavy wooden block for a base. Then, instead of closing one end of the barometer-tube used by fusion, it would be better to cement on the end a metal ferrule, into which is fitted a screw for closing the end. With such an arrangement, it will be easy to regulate the amount of air enclosed.

III. LAW OF PASCAL.

166. Apparatus. — Brass Tube, Haldat's Apparatus, Equilibrium Vases, Glass Tubing, etc.

167. Exercise. — Procure a piece of brass tubing 25 mm. in diameter and 25 cm. long. Fit to it a piston made by winding candle-wick around one end of a piece of barometer-tube 20 cm. long (Fig. 79). Close one end with a perforated cork. Drill holes through the side of the tube at different distances from the open end, and insert perforated corks. Fit neatly in all these corks glass tubes of such a length, and so bent, that their free ends are all on the same level, and at least 5 cm. above the open end of the brass tube. Remove the piston, fill the apparatus nearly full of water, and then insert the piston, holding the tube in a vertical position. Compare the heights of the water in all the tubes as pressure is applied to the piston. Why does the water rise in each tube? How does the direction of the applied pressure compare with that of the motion observed in the different tubes? What is to be learned from the comparison of the heights of the water in the tubes? Express as a proposition the truth developed by this experiment.

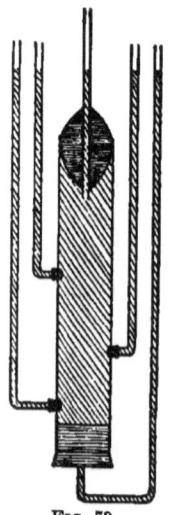

Fig. 79.

MECHANICS OF FLUIDS. 111

168. Exercise. — Bend a stout glass tube twice at right angles, making the arms 25 cm. and 12 cm. long respectively, and the distance between them 50 cm. Set this in a frame so that these arms are vertical and well supported. Take a glass tube 25 mm. diameter and 25 cm. long, a lamp-chimney of about the same length, and a two-litre bottle with the bottom removed (see p. 354); fit to each of these a cork with two holes, through which pass two pieces of tubing of the same bore, so arranged that one of them can be attached to the short arm of the U-shaped tube, and the other can be closed with a small cork, and be used as an emptying-pipe (Fig. 80). Pour mercury in the U-shaped tube till it rises 8 or 10 cm. in the long arm. Now attach one of these vessels by a short rubber connector to the U-shaped tube, and pour in water till nearly full. Mark the height of the water, and also of the mercury, in the long arm. Then replace this vessel by one of the others, and fill it with water to the same height as in the first case. In like manner, use the third vessel. What causes the mercury to rise in the long arm each time? Is there any evidence to show that it is connected with the quantity of water in the vessel attached to the short arm? Any evidence that it is connected with the height? What truth is taught by the experiment?

FIG. 80.

169. Exercise. — Cut out of wood a rectangular block 25 cm. long, 10 cm. wide, and 5 cm. thick. Bore a hole lengthwise through the stick 2 cm. in diameter. Likewise bore several holes, 5 cm. apart, through the side of the piece, so as to connect with the first one. Close the holes with perforated stoppers, rubber ones if you have them, through which pass glass tubes bent into various forms, but all terminating in a line parallel to the stick (Fig. 81). Pour water through one

FIG. 81.

of them till it rises several centimetres in that tube. Compare the level of the water in the various tubes. What causes the water to rise in the other tubes? Account for the height attained in each tube. What truth is taught?

170. Exercise. — Bore a hole in the side of a wooden pail near the bottom, and insert a stop-cock. A perforated cork, through which passes a glass tube, may be used as a substitute, by slipping over the end a short piece of rubber tubing closed with a pinch-cock. Fill the pail with water, place it on a chair on the table, and attach a piece of rubber tubing 2 or 3 metres long to this stop-cock. In the other end of this tube, insert a jet-tube secured in a vertical position by means of a support standing on the floor. Open the stop-cock, and observe closely the attending phenomenon. Explain.

Substitute for the jet-tube a glass tube of sufficient length to reach to the level of the water in the pail. Note the height to which the water rises. Account for the difference between the height reached by the water in the tube, and that when escaping as a jet. Ascertain what the height of the jet depends on, by comparing the conditions under which different heights are obtained.

IV. THE SIPHON AND PUMP.

171. Apparatus. — Glass and Rubber Tubing, several large Bottles, small Suction-Pump, Air-Pump, etc.

172. Exercise. — Bend into a U-shape a glass tube about 50 cm. long. Fill the tube with water, close one end with the finger, and invert it. Why does not the water run out of the tube? Now let each end dip into a cylindrical jar partly filled with water. Place one of the jars on a block of wood, and observe the motion of the liquid. Is the velocity of flow constant? When greatest? When least?

Make a second siphon by connecting together two L-shaped glass tubes with a rubber connector. Ascertain if the apparatus will work when a small hole is made in the side of the rubber tube between the two glass tubes. Explain.

Make a third siphon similar to the first, using tubing of about 1.5 mm. bore to diminish the rate of flow. Connect two flasks with it, place the apparatus beneath the bell-jar of an air-pump, and then quickly exhaust the air.

Examine these different experiments, and frame an explanation of the action of a siphon.

173. Exercise. — Make a siphon out of a rubber tube. Let the outer arm be a few centimetres longer than the inner one, and measure the amount of flow during five minutes.

Increase the length of the outer arm, leaving the inner arm unchanged, and again measure the flow during five minutes. Again increase the length, and measure the flow. On what does the velocity of flow depend?

174. Exercise. — Determine the law which regulates the height over which a liquid can be carried by a siphon.

Use a long rubber tube of small bore for a siphon, and mercury for the fluid. Fill the tube with mercury, and tie up the ends securely with wire. After placing the tube in position, with its ends opening in dishes having different levels, the higher containing mercury, open the ends, and gradually raise the bend of the tube till a position is reached in which the flow ceases, but is resumed on lowering it in the least. Compare the height of this above the surface of the mercury in the higher vessel, with the barometer reading at the time.

A glass siphon of small bore could be substituted for the rubber, and the experiment made under the bell-jar of an air-pump, the reading of the pump-gauge being taken at the instant the mercury-column breaks in the siphon, and this compared with the height of this column.

175. Exercise. — Devise a fountain by conducting the water from a vessel to a jet-tube by means of a siphon. On what does the height of the jet depend? Why?

176. Exercise. — Construct an intermittent fountain.

Cut off the bottom of a large bottle. See p. 354. Cork the mouth of the bottle with a perforated cork through which passes a glass tube, reaching to within 2 or 3 cm. of the end of the bottle. Support the bottle, mouth downward, in the ring of the iron stand; place a long test-tube over the glass tube,

MECHANICS OF FLUIDS. 115

such, that, when the bottle is full of water, the upper surface is above the test-tube. Arrange a siphon with an adjustable pinch-cock to convey water from some large vessel to the bottle at a rate somewhat less than the outlet-tube can empty it. Set the siphon in operation, and notice the character of the flow from the bottle. Explain.

177. Exercise. — Select three bottles of about one litre capacity each, and fit to them corks with two holes. Only the best corks can be used. Place two of the bottles on the table, and the third one on the floor. Using rubber and glass tubing, connect them as shown in Fig. 82. Now fill the bottles on the table with water, leaving the one on the floor filled with air, and set the apparatus in operation by blowing through the short tube of the farther bottle. Explain. Why must the corks be tight? Find, by varying the connections, upon what the height of the jet depends.

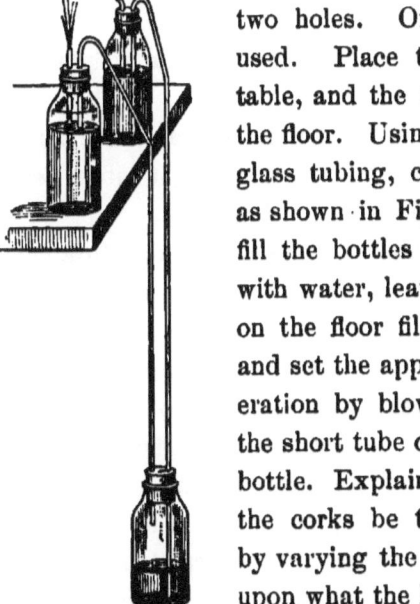

FIG. 82.

FIG. 83.

178. Exercise. — Fit a good cork to an open-top receiver. Through this cork, pass, air-tight, a glass tube reaching to the bottom of a tumbler of water within the receiver. Connect with this tube a small suction-pump (Fig. 83). Ascertain whether water can be pumped out of the tumbler. Then place the apparatus on the table of an air-pump, exhaust the air, and ascertain the effect on the working of the pump. Inference.

V. THE PRINCIPLE OF ARCHIMEDES.

179. Apparatus. — Balance, Bucket and Cylinder Apparatus, Cylindrical Graduate, Cylindrical Jar, Air-Pump, etc.

180 Exercise. — Ascertain the effect on the weight of a substance produced by an enveloping fluid.

Attach a heavy weight, stone or metal, to an accurate spring-balance, and take the reading. Lower the weight into a vessel of water, and again take the reading. Inference.

Repeat the experiment, using a block of wood. Inference. Under what condition does a substance appear to lose all of its weight?

Now place the vessel of water on the platform of a balance, and find its weight. Suspend each of the above articles in succession from the hook of the spring-balance, submerging it in the water if possible, and compare the difference in the weight of each substance caused by being placed in water, with the change produced in the weight of the vessel of water as shown by the balance. Account for the apparent loss of weight in the first two experiments.

Repeat all these experiments, substituting strong brine for the water. Compare the weight apparently lost with that apparently lost when water was used. Compare the weight apparently lost by a large piece of metal with that apparently lost by a small piece of the same substance. Compare the loss apparently suffered by a block of lead with that apparently suffered by a block of any other metal of the same size. On what does the amount of apparent loss depend?

181. Exercise. — Find the measure of the buoyant force.

Dealers in physical apparatus furnish a piece known as the **Bucket and Cylinder.** It consists of a cylindrical brass bucket, to which is accurately fitted a brass plug (Fig. 84). Hang the bucket from one arm of a balance, with the plug attached below it by means of a hook and thread. Counterpoise the two cylinders by weights in the other scale-pan. Now place a vessel of water beneath the apparatus, so that the solid cylinder will be submerged when the beam is horizontal. Why is the equilibrium destroyed? Pour water into the bucket till the equilibrium is restored. How much water is displaced by the plug? Compare this amount with that found necessary to restore equilibrium. How much weight do you find that a submerged body apparently loses?

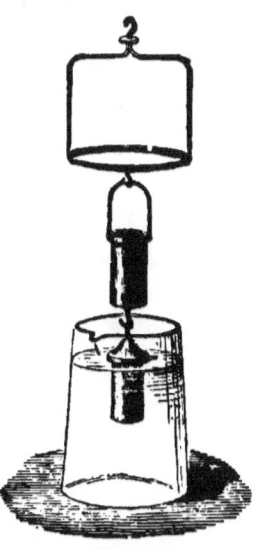

FIG. 84.

182. Exercise. — A second method of finding the measure of the buoyant force.

Tie a fine thread or hair to a piece of metal or stone, and suspend it from one end of a balance-beam. Many balances have a hook provided for attaching any substance in this way. Weigh the substance; then bring a vessel of water beneath it, and find its weight when submerged. The vessel of water can be supported on a small wooden stool placed over the scale-pan so as not to interfere with the movement of the balance. Determine how much water the substance will displace by proceeding as in Art. 20. Allowing each cubic centimetre of water to weigh a gramme, how does the weight apparently lost compare with the weight of the water displaced? State as a proposition the truth proved.

183. Exercise. — Determine the principle of flotation.

Weigh a piece of paraffine or varnished wood, then, employing a graduated cylinder containing water, determine the amount of water displaced by it when floating. If each cubic centimetre of water displaced weighs one gramme, how does the total weight of water displaced by a floating body compare with the weight of that body in air? Infer from this when a body will float in a fluid. Can the form of a body affect the displacing power? Make a piece of lead float on water without attaching any thing to it.

184. Exercise. — Weight a small test-tube, so that, when placed in a vessel of water, mouth downward, it will float, maintaining a vertical position (Fig. 85). Fill a cylindrical jar with water to within a few centimetres of the top, place in it the test-tube, and set the apparatus beneath a bell-jar on the table of the air-pump. Gradually exhaust the air, letting it re-enter the bell-jar quite frequently, to determine whether that point has yet been reached at which the test-tube is nearly on the point of sinking owing to air having been removed from it. When that point is reached, remove the jar from the air-pump, and tie over the top a piece of sheet-rubber. Now, by pressing on the top of

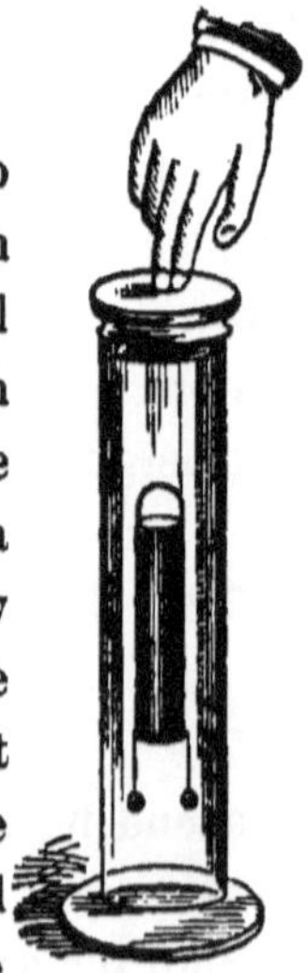

FIG. 85.

the rubber, the tube can be made to sink to any desired point within the jar. Observe the level of the water in the tube as it is made to occupy different positions by varying the pressure. Explain the action of the apparatus, pointing out all the principles illustrated by it.

In case a sufficiently tall bell-jar is not at hand, the necessary amount of air can be removed by means of a J-shaped tube, the short arm of which is about as long as the test-tube. By holding the test-tube over the short arm of the J-tube as you let it down into the water, the air will escape through this tube, water taking its place. After a few trials, it will be found easy to remove the proper amount.

185. Exercise. — Ascertain if air affects the apparent weight of substances.

Place beneath a large bell-jar on the table of an air-pump a short-beam jeweller's balance, having, in place of the scale-pans, a large cork or glass float and a piece of lead accurately balanced (Fig. 86). Exhaust the air, and note the effect on the balance. Repeat several times to make it certain that the phenomenon attending the exhaustion is not accidental. Explain. How can the weight of an object obtained in air be reduced to that in a vacuum?

FIG. 86.

VI. DETERMINATION OF DENSITY.

186. Apparatus. — Balance, Cylindrical Graduate, Specific-gravity Bottle, Hydrometers, Florence Flask, Air-Pump, Glass Tubing, etc.

187. Exercise. — Determine, by means of a balance, the density of a solid insoluble in water.

Suspend the substance, by means of a fine thread or hair, from the hook on the bail of the scale-pan provided for that purpose, and determine carefully its weight. As in Art. 182, find its weight in water. From these data determine the volume of water displaced by the solid. Divide the weight in the air by the volume, and the density is obtained. This result needs correcting for temperature, pressure, etc., when great accuracy is required.

As a check, divide the weight in air expressed in grammes by the volume in cubic centimetres, as determined by the method of Art. 20.

Tabulate results as follows: —

Name of Substance.	Weight in Air.	Volume, ccm.	Density.	Weight in Water.	Density.	Average.
Glass . . .						
Iron . . .						
Lead . . .						
Marble . .						
Copper . .						
etc.						

MECHANICS OF FLUIDS. 121

188. Exercise. — Determine the density of a substance lighter than water.

As in the last experiment, determine the weight in air, and also the weight of a piece of lead or other metal of sufficient mass to sink the specimen in water. Then determine the weight of the sinker in water, and also of the specimen and sinker combined. From these data, the volume of the water displaced by the substance is easily obtained. How? Divide the weight in air by the volume, and the required density is obtained.

Check this in the same manner as in the last experiment.

189. Exercise. — Determine the density of a substance soluble in water.

As in Art. 182, find the weight of an equal volume of some liquid of known density in which it is insoluble. This divided by the density of that liquid will give the volume of the substance. Why? Now proceed as in Art. 187.

Rock-salt is insoluble in naphtha, rock-candy and saltpetre in strong alcohol, alum and blue vitriol in oil of turpentine, etc.

190. Exercise. — Determine the density of a liquid.

For accurate work, use the specific-gravity bottle. It is a small flask holding usually 10, 25, 50, or 100 ccm., closed with an accurately fitting glass stopper, through which is a small capillary opening. Thoroughly clean the bottle, rinsing it out finally with strong alcohol, and dry by forcing into it a stream of air passing through a heated tube. After weighing the bottle, ascertain its capacity if not already known, then fill it with the liquid whose density is sought, wiping the exterior of the bottle dry, and removing the excess of the liquid that escapes through the hole in the stopper, and find its weight.

Deducting the weight of the empty bottle, divide the remainder by the volume of the bottle, and you have the density. This result, when great accuracy is required, must be corrected for temperature, etc. Consult Stewart and Gee's "Practical Physics," vol. i.

For approximate work, a small Florence flask may be used, filling it with water up to a line marked on the neck.

191. Exercise. — Determine the density of a liquid by weighing a solid in it.

Make a sinker out of a thick piece of glass rod. Fasten it by a fine thread to the hook on the frame of the scale-pan. Find the apparent loss of weight of the sinker in water, and in the liquid whose density is sought. What does this apparent loss represent? Divide the apparent loss in the liquid by the apparent loss in water, and the quotient will be the density.

192. Exercise. — Determine the density of a substance in a state of powder.

Use the flask of Art. 190, find its weight, then re-weigh it after filling it part full of the powder. Now cover it with water, place it beneath the bell-jar of the air-pump, and exhaust the air, the water taking the place of the air within the powder. If the substance is soluble in water, some other liquid must be used. On removing the flask from the pump, ascertain the weight after filling it full of water. Let f = weight of the flask, w = weight of substance after deducting weight of the flask, w' = weight of substance when covered with water, w'' = weight of water required to fill the flask when the substance is removed. Then $w' - w$ = weight of water added to fill the flask when the substance was in it, and

MECHANICS OF FLUIDS.   123

$w'' - (w' - w) =$ weight of water whose volume is that of the substance. Therefore the density $= \dfrac{w}{w'' - w' + w}$.

193. Exercise. — Determine the density of a solid with Nicholson's hydrometer.

This instrument usually consists of a metal cylinder with conical ends, to the vertices of which are soldered stout wires. On the end of one wire is soldered a small scale-pan, and on the other a perforated pan (Fig. 87). The lower end of the cylinder is loaded with shot so that, when the apparatus is placed in water, it takes a vertical position with the cylindrical part nearly submerged.

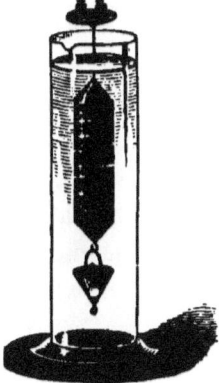

Fig. 87.

To use this apparatus, place on the upper pan a sufficient number of known weights to sink it to a mark on the stem just below the scale-pan. Let a represent the number of grammes required. Then place on the pan the substance whose density is required, and add weights till the instrument sinks to the mark on the stem. Let b represent the number of grammes added. Then $a - b =$ the weight of the substance. Now transfer the substance to the lower pan, tying it on if necessary, and observe how many weights must be added to the b weights to sink the instrument to the mark on the stem. Let c represent the number of grammes necessary. Then $\dfrac{a-b}{c} =$ density. Why?

194. Exercise. — Determine the density of a substance with the Jolly balance.

Solids are weighed as directed in Art. 24, but with the lower pan attached, and immersed in water. The weight in water is obtained by placing the substance on the lower pan.

To find the density of a liquid, weigh a solid insoluble in it, first on the upper pan, then on the lower pan when immersed in water, and finally on the lower pan when immersed in the liquid. Then proceed as in Art. 191.

195. Exercise. — Determine the density of a liquid by means of a Nicholson's hydrometer.

Weigh the instrument, then place it in water, adding weights to the upper pan till it sinks to the mark on the stem. The weight of the water displaced by the instrument will be the sum of the weights on the pan and that of the instrument. Why? In like manner, determine the weight of the liquid displaced, whose density is required. From these data, the density is easily determined. How?

For corrosive liquids, a glass hydrometer must be used. In such instruments, the lower pan is usually omitted.

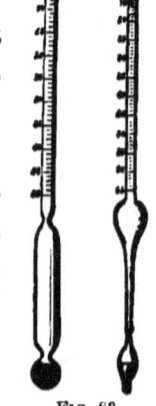

Fig. 88.

196. Exercise. — Make out of light wood a bar of uniform size throughout, having a length of 30 cm. and a cross-section of 1 scm. Bore a hole several centimetres deep into one end, and put in pieces of lead sufficient to maintain the bar in a vertical position when floating in water, closing the hole with cement. Lay off on the bar a decimal scale, the zero at the loaded end. Give the apparatus a thin coat of paraffine. Now float the instrument in a vessel of water, and note the division of the scale to which it sinks; then place it in the liquid whose density is sought, and take the

MECHANICS OF FLUIDS.

reading. The quotient of the former by the latter is the density of the liquid. Why?

For accurate determinations, especially of corrosive liquids, glass hydrometers (Fig. 88) with graduated stems are usually employed. The method of use is similar to that illustrated above.

Make a five-per-cent, a ten-per-cent, and a twenty-per-cent solution of common salt, and determine, by means of the hydrometer of constant weight, the density of each. Verify the results by some one of the methods already given.

197. Exercise. — Fit a rubber stopper with two holes to a small Florence flask. In this stopper, insert, air-tight, two glass tubes, each 75 cm. long, and bent twice at right angles at the end next the cork (Fig. 89), so that the tubes will not stand so close together. Support the apparatus in a vertical position, with the end of one tube dipping into a beaker of water, and the other into a vessel containing the liquid whose density is sought. Carefully measure the capillary effect. Now gently warm the flask till part of the air is expelled. When cold, the liquids having ceased to rise, measure the height of each liquid above that in the vessel. The ratio of these heights corrected for capillarity will be the inverse ratio of their densities.

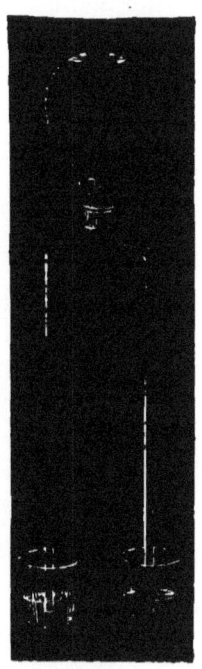

FIG. 89.

198. Exercise. — Determine the density of air.

Fit to a stout Florence flask of about one litre capacity a rubber stopper through which is inserted a good stop-cock.

Dry the flask, and determine its weight. Then exhaust the air as completely as possible, reading the pressure-gauge on the pump as well as the barometer. Weigh the exhausted flask, then measure its capacity, and compute from these data the weight of one cubic centimetre of air, that is, its density. The capacity of the flask can be found by weighing it full of water.

For more accurate methods, consult Roscoe and Schorlemmer's "Chemistry," vol. iii. pt. i.

HEAT. 127

CHAPTER IV.

HEAT.

I. HEAT, AND MECHANICAL MOTION.

199. Apparatus. — Flint and Steel, Fire-Syringe, Tinder, Bottles, Thermometer, Metal Rod, Tumblers, Mercury, Whirling-Machine, Air-Thermometer, Condensing-Pump, Air-Pump, Florence Flasks, Pyrometer, Radiometer, etc.

200. Exercise. — Hold a piece of hardened steel, as the back of the blade of a large pocket-knife, between the thumb and forefinger of the right hand, and strike a glancing blow against the sharp edge of a piece of flint. Observe any attending phenomena. Examine the substances after striking them together a few times. Is there any evidence of any quantitative connection between the energy expended and the heat obtained?

201. Exercise. — Place a small piece of German tinder in the cavity at the end of the piston of a Fire-Syringe (Fig. 90), and force it quickly into the cylinder. Remove the piston as soon as possible, and examine the tinder. After each introduction of the piston, fresh air must be introduced into the tube to supply oxygen.

FIG. 90.

If the syringe has a glass cylinder, introduce within it a pellet of cotton-wool moistened with carbon disulphide.

Remove it after it has remained a few moments, and force in the piston, keeping a close watch on the tube.

Account for the phenomenon attending this experiment.

202. Exercise. — Fill a bottle of about four ounces capacity half full of water, and take the temperature with a thermometer. The water should be at the temperature of the room. Now cork the bottle, wrap it up in several thicknesses of paper to keep heat from being communicated to it from the hand, shake vigorously for an observed number of minutes, and again take the temperature. Ascertain the effect of increasing the time of shaking. Is there any evidence of any connection between the amount of motion and the heat? Devise some way of showing that the rise in temperature was not due to heat communicated to the water from the hand.

203. Exercise. — Rub a metal rod vigorously with a piece of leather for a few minutes, protecting the rod from the heat of the hand by a paper holder. Touch the rod to a piece of phosphorus. Inference. Have you any equivalent for the energy expended in rubbing the rod? Is there any evidence of any quantitative relation between the heat developed and the energy expended? If energy is indestructible, what would you infer is the nature of heat?

Caution. — Always cut phosphorus under water, and never handle it with dry fingers.

204. Exercise. — Wrap each of two tumblers with several thicknesses of paper to keep from them the heat of the hand. Fill one of the tumblers one-third full of mercury, and determine its temperature by inserting a thermometer. Then pour the mercury rapidly, in a stream as large as a slate-pencil, back and forth between the tumblers for five times, letting the mercury fall from a height of about 25 cm. Take the temperature. Ascertain the effect of ten pourings; twenty. Repeat these tests, letting the mercury fall 50 cm. Inference.

The mercury should be at the temperature of the room at the beginning of each set of pourings.

205. Exercise. — Attach to a whirling-machine a brass tube about 15 cm. long and 15 mm. diameter (Fig. 91). Fill it half full of alcohol, and close the end with a cork. Rotate

FIG. 91.

the apparatus rapidly, applying a brake made of two pieces of board hinged at one end with a strip of leather, and also covered with leather at the places of contact with the tube. In a few minutes, the alcohol can be made to boil vigorously, as the experimenter will find ample evidence. Account for the heat.

206. Exercise. — Blow a stream of air with a small bellows against the bulb of a sensitive thermometer, or the face of a thermopile (Fig. 114), and record the effect. Explain.

207. Exercise. — Compress air within the copper reservoir of the condensing-pump, obtaining as high a pressure as possible. After the apparatus has stood long enough to acquire the temperature of the room, turn the stop-cock, directing the escaping air against the bulb of a thermometer, or the face of the thermopile, and record the effect. Whence does the air, on escaping, acquire its kinetic energy? Devise some way of proving your answer.

Suspend a thermometer within the receiver of the air-pump, observing the effect as the exhaustion proceeds. Explain.

208. Exercise. — Fit, air-tight, to a Florence flask a cork through which passes a J-shaped tube reaching nearly to the bottom of the flask, and a thermometer with the bulb at the centre of the flask (Fig. 92). Fill the flask a little less than half full of water at the temperature of the room, and hold it submerged in a vessel of boiling water for 30 seconds. Note carefully the effect produced on both the water of the flask and the thermometer. Now remove the J-tube, and close, air-tight, the opening in the cork. Change the water in the flask for an equal amount of water of the temperature of the room. Again hold the flask, for 30 seconds, in a vessel of boiling water, and record the reading of the thermometer. Account for the difference of temperature in the two cases.

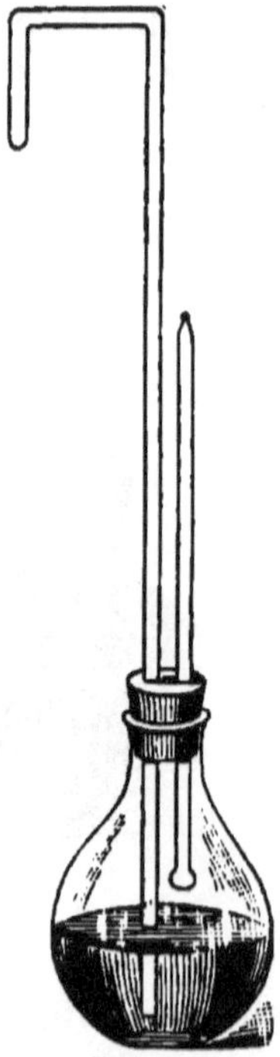

FIG. 92.

HEAT.

It will be necessary to hold the cork in the flask, or the elastic force of the air will throw it out, and break the thermometer.

209. Exercise. — Fasten a metal rod about 30 cm. long to a wooden block by means of small wire staples, the block to be attached to a board for a base. To the opposite end of the board, tack a thin piece, to support an index against which rests the free end of the rod (Fig. 93). The pointer may be cut out of tin or brass, and should have the pivot distant but a couple of millimetres from the end of the rod. Place a spirit or gas flame under the rod for a few minutes, and watch the effect.

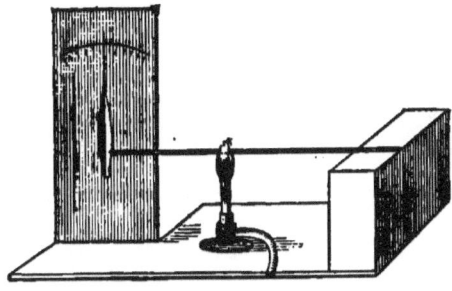
Fig. 93.

Fasten the end of the pointer, by an elastic, to a pin, and see if the expanding rod does work by stretching the elastic. Whence the energy expended in doing this work?

210. Exercise. — Place a Radiometer (Fig. 94) in a beam of sunlight, or in the light from a gas-jet admitted through an opening cut in a large sheet of cardboard, and notice the effect on the instrument. Try a Bunsen or spirit flame which emits heat and but little light. Interpose between the radiometer

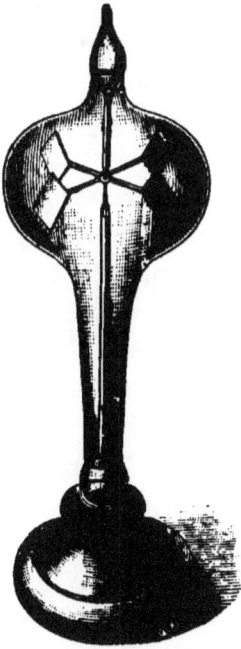
Fig. 94.

and the source of light a chemical tank (Fig. 234) or flat bottle filled with alum-water, a substance which cuts off but little light and most of the heat. Substitute a solution of iodine in carbon disulphide in place of the alum-water, thus cutting off the light, and but little of the heat. Inference.

II. HEAT, AND CHEMICAL ACTION.

211. Apparatus. — Test-Tube, Thermometer, Saucer, Mortar, etc.

212. Exercise. — Fill a test-tube one-third full of water, and take its temperature with a thermometer. Fill a second one one-third full of sulphuric acid, and likewise take its temperature. Now pour the acid in a fine stream into the water, and record the temperature. Inference.

213. Exercise. — Place in a common saucer a lump of freshly burned quicklime as large as a hen's egg. Ascertain its temperature by placing the bulb of the thermometer in contact with it for a few minutes. Now fill the saucer with water, and, after a few minutes, take the temperature of the lime. Explain.

214. Exercise. — Powder in a mortar some crystals of cupric nitrate. Spread the powder thickly on a piece of tin-foil about 10 cm. square. Sprinkle on the powder a small quantity of water, and quickly wrap it up in the foil, pressing down the edges. Notice the changes which follow. Explain.

III. CONDUCTION OF HEAT.

215. Apparatus. — Thermometer, Conductometer, Rods of different kinds of material, Cylinder of Brass, and one of Wood, Lead Balls, Brass Wire-Cloth, Wooden Cube, Air-Thermometer, etc.

216. Exercise. — Take the temperature of several different articles in a room, at about the same distance from the floor, by holding the bulb of the thermometer in contact with each successively for a few minutes. Now test the same substances as to temperature by touching them with the hand. Which substances feel cold, and which do not? Account for the differences between the conclusions based on the sense of touch, and those on the indications of the thermometer.

217. Exercise. — Compare the conductivity of several substances.

Get a tin-smith to construct a tin dish with small tubes on the side, in which can be cemented rods of different substances, as iron, copper, zinc, brass, etc. (Fig. 95). Coat these rods with a thin layer of beeswax; then fill the tank with hot water, and compare the rates at which the wax melts on the rods.

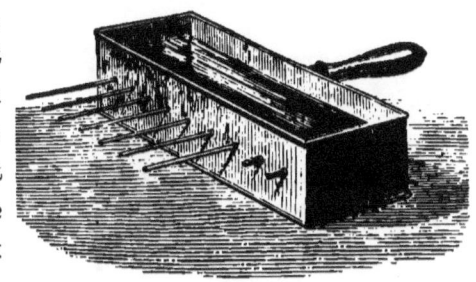

FIG. 95.

As the four substances mentioned differ but little in their capacity for heat, the rates of transference of temperature will not differ much from those for transference of heat.

218. Exercise. — Wrap firmly around a brass cylinder a layer of writing-paper, hold it in the flame of a lamp, and record the time required to set the paper on fire. Then wrap a similar piece around a wooden cylinder of the same size, and hold it in the flame till the paper burns. Account for the difference in the times.

219. Exercise. — Compare the thermal conductivity of iron and copper.

Select a stout wire of each, 35 cm. long, twist them together for about 10 cm., and then insert the twisted end through a wooden frame, the untwisted parts being separated (Fig. 96). Attach lead balls or marbles to the wires at regular intervals between the supports, using shoemaker's wax for the purpose. Place a spirit-lamp under the twisted part, and record the effect on the balls. Inference.

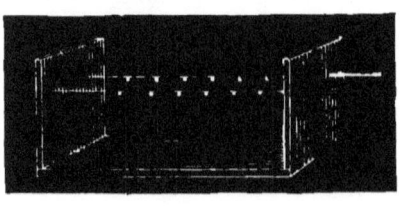

Fig. 96.

220. Exercise. — Support a piece of brass wire-cloth on the ring of the iron stand, place a gas-flame beneath it, and record the effect on the flame (Fig. 97). Test for gas above the gauze by applying a lighted match. Account for the flame's not extending through the gauze at first. Now let the wire-cloth cool, then turn on the gas, and light it above the cloth.

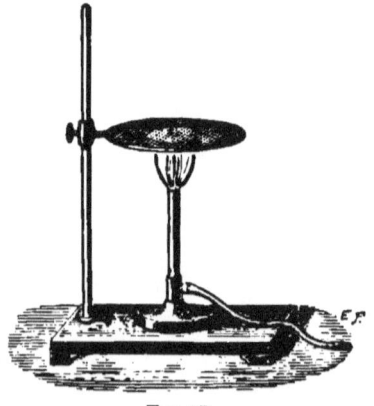

Fig. 97.

Account for its burning above the wire-cloth, and not below. Observe the manner in which gum-camphor burns when placed on the wire-cloth.

221. Exercise. — Compare the conductivity of wood with the grain, with that across the grain.

Coat uniformly with paraffine three of the faces about one corner of a wooden cube 5 cm. on the edge. Heat a metal ball of 2.5 or 3 cm. diameter in boiling water, and place it as quickly as possible on one of the waxed faces of the cube. Compare the shapes of the melted spots formed on the different sides, and account for the difference. The ball must remain on the block the same length of time in each case. Try cubes made from pine, oak, maple, etc.

Wipe the water off of the ball before placing it on the wax.

222. Exercise. — Test the conductivity of water.

Place in a long test-tube some fragments of ice. Fill the tube nearly full of water, keeping the ice from floating by means of a small marble or stone. Hold the tube in an inclined position in a Bunsen or spirit flame, heating the part near the surface. Compare the temperature of the water at the bottom with that at the top on its coming to a boil. Inference.

In heating the tube, pass it slowly backward and forward through the flame, as a sudden application of heat is liable to break the tube. Why?

223. Exercise. — Cork the neck of a large funnel, and through it pass the stem of an air-thermometer, bringing the bulb below the surface of the funnel. Support the apparatus

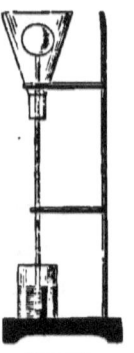

Fig. 98.

as shown in Fig. 98. Fill the funnel with water till the bulb of the air-thermometer is covered to the depth of a centimetre. Pour a spoonful of ether on the water, set it on fire, and record the effect on the thermometer. Inference.

The lower end of the thermometer-tube should dip into colored water. By warming the bulb with a gas-flame before introducing water into the funnel, some of the air can be expelled from the bulb, so that, on cooling, water will stand at an elevation of a few centimetres in the tube.

IV. CONVECTION OF HEAT.

224. Apparatus. — Florence Flasks, Glass Tubing, Lamp-Chimneys, tall Bell-Jar, large Bottle, etc.

225. Exercise. — Fill a large Florence flask two-thirds full of water, introduce a little powdered cochineal or paper raspings, and apply heat by placing it on a sand-bath on the iron stand. Make a careful study of the action of the particles, to ascertain if any law governs their movements. Explain.

226. Exercise. — Construct out of glass tubing a rectangle about 12 cm. long by 50 cm. wide, making the connections with rubber tubing. A T-tube should be inserted at one corner for convenience in filling, and to provide for the expansion of the water. Fill the apparatus with water freed from air by boiling, and introduce a little powdered cochineal. Now hold one corner of the apparatus in a vessel of hot water, and

HEAT. 137

observe the effect on the water in the tube. Apply ice or snow to the upper corner of the rectangle, and record the effect. Make the application at other points. Account for the results.

227. Exercise. — Support a short piece of candle in a vertical position in the bottom of a common plate. Light the candle, and place over it a wide cylindrical lamp-chimney. Pour water into the plate to prevent air entering the chimney at the bottom. Note the effect on the flame. Insert a cardboard partition in the chimney, reaching nearly to the flame, and note the effect. Soak some porous paper in a solution of saltpetre, and dry it thoroughly. This is known as touch-paper, and burns with the emission of considerable smoke. Hold a piece of burning touch-paper over the top of the chimney, and make a study of the air circulation.

228. Exercise. — Paste paper over the cracks of a common chalk-box to make it air-tight. Near one end of the cover, make a number of small holes in a circle somewhat smaller than the base of a lamp-chimney (Fig. 99). In the centre of this ring, place a lighted taper. Near the other end of the cover, cut a hole 25 mm. in diameter. Place a lamp-chimney over each of these openings, coating the edge of each with paraffine to produce an air-tight joint. Study the air currents through the apparatus by holding burning touch-paper over the tops of the chimneys. Ascertain the effect of closing the top of the chimney not containing the taper.

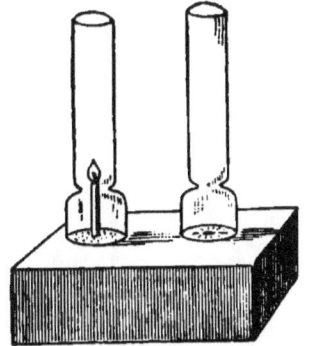

FIG. 99.

229. Exercise. — To a tall wide-mouthed jar, fit a cork ring supporting a wide glass chimney. Place the jar over a candle standing in a plate of water. After observing the effect on the flame, insert in the chimney a smaller one extending down nearly to the flame, supporting it by a wire frame (Fig. 100). Study the air-currents with touch-paper.

Paper cylinders may be substituted for the chimneys.

230. Exercise. — Invert a tall jar, at least 50 cm. high, over a stand supporting several tapers (Fig. 101). The stand may be made by inserting a stout wire into a circular piece of board for a base, supporting the tapers on wire arms made to retain their places on the standard by the friction of the spiral formed on one end. Observe the order in which the tapers are extinguished. Why do the tapers go out? What is implied by the order in which they are extinguished? Ascertain the effect of introducing fresh air by means of a tube reaching to the top of the standard, and connected to a bellows.

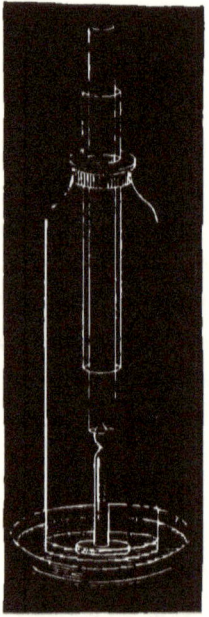

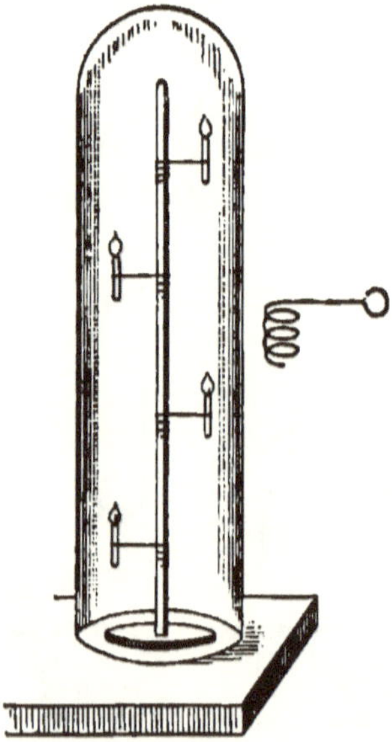

Fig. 100.

Fig. 101.

HEAT.

231. Exercise. — Test different parts of the schoolroom for carbonic acid.

Prepare a number of litres of lime-water by stirring quicklime in distilled water, and, after standing a few hours, filtering it through porous paper (Fig. 102). Select a clear white glass bottle, holding, when full, $10\frac{1}{2}$ fluid ounces of water. Dry it thoroughly, and fill it with the air to be tested by taking the bottle to that part of the room, and sucking out the air in it by means of a glass tube. Avoid breathing through the tube into the bottle. Now introduce half a fluid ounce of clear lime-water, cork the bottle, and shake vigorously for several seconds. If, after the air-bubbles have disappeared, the liquid is not clear, the air examined contains at least 0.06 per cent of carbonic acid. If an eight-ounce bottle is used, turbidity indicates at least 0.08 per cent. A six-and-a-half-ounce bottle will show 0.1 per cent; a five-and-a-half-ounce bottle will show 0.12 per cent. If one ounce of lime-water is used, a seven ounce bottle will show 0.2 per cent; a five-ounce bottle will show 0.3 per cent; a four-ounce bottle will show 0.4 per cent. Compare the air of a room which has recently been occupied by a large class with that of one which has been vacant for several hours. Test the air at the top, at the middle, and at the bottom of the room.

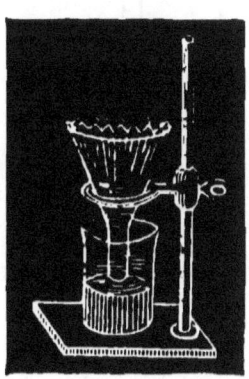

FIG. 102.

Breathe through a glass tube, holding the end in a vessel of lime-water, and note the effect. Also let a candle burn for a short time in a large bottle, then remove it, pour in lime-water, shake the bottle, and mark the effect. Account for the air of rooms becoming vitiated.

V. EXPANSION BY HEAT.

232. Apparatus. — Heavy Wire, Pyrometer, Metal Ball, Caliper, Compound Bar, Florence Flasks, Glass Tubing, Thermometer, etc.

233. Exercise. — Ascertain whether heat affects the length of a wire.

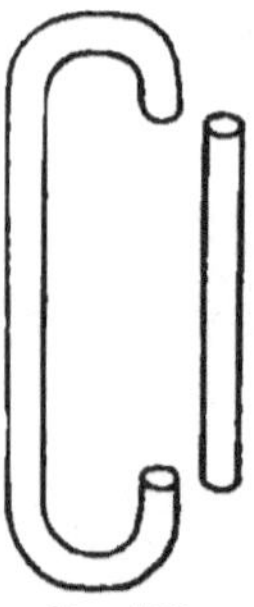

Bend a thick brass or iron wire about 25 cm. long into a rectangular figure; cut out of one side a piece 5 cm. long, and adjust the link so that this piece fits exactly the place from which it was taken when at the temperature of the room (Fig. 103). The ends of the parts must be planes perpendicular to the axis of the wire. Hold the straight piece with a pair of tongs in a flame, and, when hot, try to insert it in its place in the link. Inference.

Fig. 103.

234. Exercise. — Determine the coefficient of expansion of a metallic rod.

Employ an apparatus of the form shown in Fig. 104, in which, by means of a compound lever, the expansion is magnified. HK is a copper tank 30 cm. long, 5 cm. deep, and 3 cm. wide. This rests on the base of the instrument, and is prevented from moving towards K by a brace L. In this tank, place the rod, rest it on short pieces of glass tubing placed across the bottom. One end of the rod must press firmly against the end K of the tank; the other end must press against the flat vertical rod D which is soldered at right angles to the rod BF supported by the block P, and free to turn about its axis without play. The bearing at A is made in two parts, and works in a groove in the rod BF, preventing

HEAT.   141

motion lengthwise. The pointer CE is soldered at right angles to BF at B, the part BE being heavy enough to cause D to press firmly against the end of the rod in the tank. The expansion of the rod will cause the pointer to move over the scale M, and the space traversed divided by the ratio of BC to the arm D will be the expansion of the rod. It should be

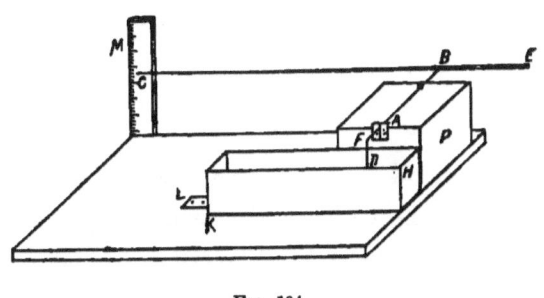

FIG. 104.

observed that the arm D extends from the centre of the end of BF to the point of contact of the upper edge of the rod in the tank with D.

To use the apparatus, measure accurately the rod by laying it on a finely divided steel scale, then place it in the tank and record the reading of the pointer and the temperature of the rod. Now fill the tank with boiling water, and again record the reading of the pointer and the temperature of the rod or water. From these data compute the rod's change in length, and also its change per unit of length for one degree.

235. Exercise. — Determine the effect of heat on a metal ball.

Adjust a pair of calipers so that an iron or brass ball will just pass between the points. Again caliper the ball after it has stood for a few minutes in hot water. What fact regarding the effect of heat on substances is shown by this experiment different from that shown in the last experiment? Compute the change in volume per degree from the change in diameter.

236. Exercise. — Select two pieces of brass tubing about 8 cm. long each, such that one will just telescope the other. Fit to each a wooden handle. Heat the smaller tube in a gas-flame, and try to insert it within the larger one. Heat both tubes, and then try to insert one within the other. Inference.

237. Exercise. — Fasten together by rivets, at intervals of 2 cm., a strip of sheet-iron, and one of copper or brass, each 15 cm. long and 2 cm. wide. Place this bar on the supports A and B of the apparatus shown in Fig. 36. Place a spirit-lamp beneath the centre of the bar. What is implied by the movement of the indicator? Ascertain the effect of placing ice on the upper side of the compound bar.

238. Exercise. — Determine the effect of heat on the volume of a liquid.

Fit a perforated cork to a Florence flask, and insert a glass tube about 30 cm. long, just reaching through the cork. Fill the flask and part of the tube full of water, excluding all air-bubbles. The cork and tube must fit water-tight. Place the apparatus on the sand-bath of the iron stand. Mark the position of the water in the tube. Apply heat, and note the effect. Inference.

239. Exercise. — Determine the coefficient of expansion of a liquid.

HEAT.   143

Close one end of a glass tube 15 cm. long and about 5 mm. in diameter, by heating the end in a gas-flame. Select one of uniform bore. Fill the tube part full of some liquid, as water, alcohol, glycerine, ether, or naphtha, and then tie it to a chemical thermometer, so that the thermometer-scale may be used in measuring the liquid in the tube. Support the apparatus in a beaker of ice-water, and, when it has acquired the temperature of the water, ascertain the length of the liquid column in terms of the spaces of the attached thermometer-scale. Then plunge the apparatus into a vessel of hot water, and note the thermometer reading, and the length of the liquid column after the liquid has ceased rising in the tube. The part of the tube containing the liquid must be kept in the water so that all the liquid may be at the same temperature. Compute, from the data, the expansion per unit of volume for one degree, neglecting the effect of the heat on the capacity of the tube.

240. Exercise. — Determine the effect of heat on the volume of air.

Procure a heavy glass tube about 30 cm. long and 2 mm. internal diameter, one end terminating in a bulb of about 5 cm. diameter. A small Florence flask closed with a good cork, through which passes a glass tube, will make a fair substitute. Support the tube in a vertical position, with the end dipping into a shallow vessel of colored water. Apply heat to the bulb by either a gas-flame, or by pouring on hot water. Note the effect. Inference.

241. Exercise. — Determine the coefficient of expansion of air.

Procure a glass tube about 20 cm. long and 1 mm. internal diameter. Introduce into it an index of mercury 5 mm. long by plunging it into a bottle of mercury, and then holding the finger firmly over the outer end of the tube as you remove it from the bottle. By inclining the tube, move the index along to the middle of the tube, and then close one end of the tube by holding it, in a horizontal position, in a gas or alcohol flame. Tie the tube to a chemical thermometer so that the thermometer-scale may be used as one of equal parts in measuring the length of the confined air-column in the tube. Place the apparatus in melting ice, and determine the length of the air-column when the thermometer-reading is zero. Before taking the readings, tap the tube with the finger to facilitate the movement of the mercury index. Now place the apparatus in a vessel of water so arranged that the temperature can be increased at pleasure, and determine the length of the air-column for 10° C., 20° C., 30° C., etc. A long shallow dish should be employed in order that the apparatus may be placed in nearly a horizontal position to remove the pressure on the air produced by the index. Such a dish can be easily made out of sheet-lead, closing the joints with plaster-of-Paris. Heat the tube by pouring hot water into the dish. Compute, from the data, the expansion per degree of the air per unit of volume at 0° C., between 0 degree and 10 degrees, 10 degrees and 20 degrees, 20 degrees and 30 degrees, etc., neglecting the errors arising from the effect of heat on the volume of the tube, and from the lack of uniformity in the bore of the tube.

242. Exercise. — Fit a perforated cork, through which passes a glass tube, to a test-tube. As a test-tube will stand but little pressure, soft and elastic corks must be employed. Fill the tube with water, recently boiled to expel the air, half-way up the inserted tube. Pack the apparatus in finely broken ice, and watch the water-column for some time. Remove from the ice, and, with as little delay as possible, plunge the tube into hot water, recording the effect on the column. Inference.

243. Exercise. — Close one end of a piece of lead pipe about 30 cm. long and 15 mm. in diameter by flattening it, on an anvil, by a blow from a hammer. To the other end, fit a perforated cork, through which passes a glass tube about 20 cm. long. Bend the lead into a coil, the plane of which is perpendicular to the glass tube (Fig. 105). Now fill the apparatus with water till it stands at a height of 4 or 5 cm. in the glass tube. Care must be taken to exclude all air-bubbles from the coil. Then pack the apparatus in a freezing mixture made of three parts of pounded ice, and one of common salt. Mark the level of the water in the tube when first placed in the freezing mixture. Observe the change in level during the following twenty minutes. Explain.

Fig. 105.

244. Exercise. — Prepare an apparatus similar to that used in Art. 242. Introduce several pieces of dry ice, then fill the apparatus with kerosene till the column reaches within a few centimetres of the top of the tube. Observe the change of level as the ice melts. Measure the amount of change of volume; measure the amount of water produced by the ice. From these data, compute the density of ice.

VI. THERMOMETRY.

245. Apparatus. — Thermometer, Funnel, Florence Flasks, Air-Pump, Condenser, Test-Tubes, etc.

246. Exercise. — Test the accuracy of the location of the freezing and the boiling points on a thermometer-stem.

To ascertain if the zero-point is accurately located, support the thermometer in a burette-holder, and pack around the bulb pounded ice as far up the stem as the zero-point. The vessel containing the ice should have a hole in the bottom for drainage. A common funnel will answer (Fig. 106). Wash the ice thoroughly before using, so that the melting-point will not be interfered with by the presence of foreign substances. After the thermometer has remained a few minutes in the melting ice, observe the reading of the thermometer, using a magnifying-glass to estimate fractions of a degree.

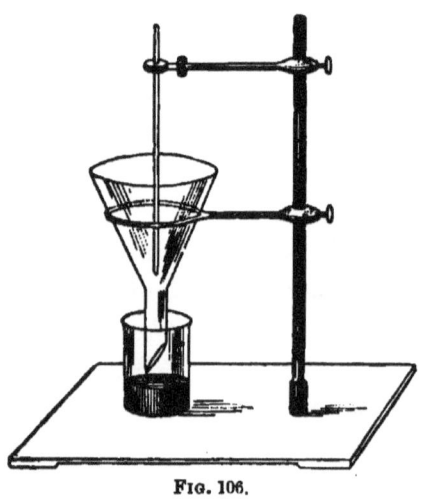

FIG. 106.

To test the accuracy of the location of the boiling-point, fit into a wide-mouthed Florence flask a large glass tube, as a lamp-chimney, placing candle-wick around the base of the tube, so that steam from the water in the flask escapes only from the top (Fig. 107). Hang the thermometer within this tube, with the bulb 2 or 3 cm. above the water in the flask, and the boiling-point on the stem just above the top of the tube. On boiling the water in the flask, the mercury-column is enveloped in steam, and will be heated quite uniformly. After the lapse of several minutes, the apparatus having time to become thoroughly heated, read the height of the mercury-column, and also observe the barometric pressure. As the boiling-point is affected by changes in the atmospheric pressure, to obtain the true boiling-point for the pressure at the time of the experiment, apply the formula $t = 100° + 0°.0375 (b - 760)$, in which b is the observed barometric reading expressed in millimetres.

The record may be kept as follows: —

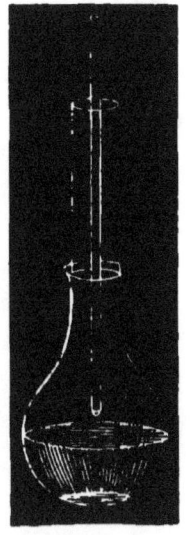

FIG. 107.

Freezing-point. . . . . Error of instrument . . ..
True boiling-point . . .., as computed for pressure of .... mm.
Boiling-point observed . ... Error of instrument . . ..

247. Exercise. — Compare the temperature of boiling water with that of the steam given off.

Boil some water in a flask. Compare the reading of the thermometer when suspended within the flask, with the bulb about two centimetres above the boiling water, with that when the bulb is in the water. Ascertain the effect of placing rough pieces of glass or stone in the flask. Account for the difference.

248. Exercise. — As in Art. 246', locate the freezing-point on the thermometer-stem. Now hold the thermometer for some moments in boiling water, and then replace it in the vessel containing the melting ice, observing again the freezing-point. Account for the difference.

249. Exercise. — Fill a vessel of about one litre capacity half full of cold water. Pour into it nearly as much more boiling water, letting the stream strike a floating cork to break the fall. Then observe the temperature at the top, at the bottom, and midway. Now stir the water with the thermometer, and again observe the temperature. Account for the differences. What precaution does this experiment suggest should be taken in obtaining the true temperature of a liquid?

250. Exercise. — Assuming that one thermometer is correct, compare a second one with it, and make out a table of corrections to be applied to its readings.

Suspend the two instruments side by side with the two bulbs on the same level. Pack the bulbs in melting ice, and record the reading of each. On removing the ice, place beneath the thermometers a beaker of water so arranged that its temperature can be varied at pleasure by introducing ice or applying heat. Beginning with ice-water, apply heat, stirring constantly with a glass rod, and record the readings of the thermometers, as they hang with their bulbs immersed in the water, for every change of 10 degrees in the standard, till the boiling-point is reached. Conclude the observations by returning the standard to the melting ice, and recording the reading. Correct the standard by the deviation of this last reading from zero. Reduce all the readings to the same thermometric scale, and obtain the differences between corresponding ones. These

HEAT.

differences will be the corrections to be applied to the thermometer under examination to make its readings standard.

Enter results as follows: —

Standard . . . . 0°, 10°, 20°, 30°, 40°, 50°, 60°, 70°, 80°, etc.
Standard corrected . . . .
The one being tested . . . .
Reduced to scale of standard, ...
Corrections to be applied . ...

251. Exercise. — Determine the general effect of changes in the atmospheric pressure on the boiling-point.

Fill a small beaker half full of water, and heat it over a lamp till it nearly reaches the boiling-point. Take the temperature, and then place the beaker under the receiver of an air-pump. As you exhaust the air, note from time to time the temperature of the boiling water as shown by a thermometer suspended in it, at the same time reading the pressure-gauge of the pump. Inference.

Try alcohol, ether, naphtha, etc.

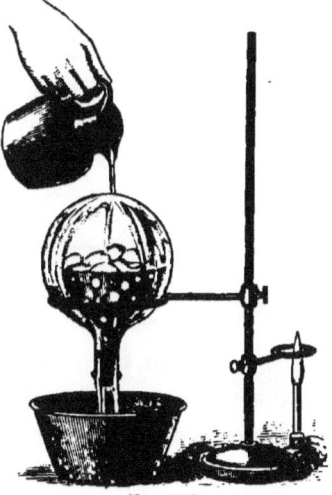

252. Exercise. — Fill a round-bottomed Florence flask half full of water, and heat it over a lamp. After the boiling has continued for

FIG. 108.

some minutes, and the air has been expelled, remove the flask from over the lamp, cork tightly, support it in an inverted position (Fig. 108) on a ring of the iron stand, and pour over it cold water. Account for the result.

253. Exercise. — Determine the melting-points of such substances as tallow, lard, paraffine, beeswax, etc.

Fill, by suction, with the melted substance, a capillary glass tube having thin walls, and close the end by fusing it in a lamp. On cooling, you will have a fine opaque thread of the substance in the tube. Fasten this tube alongside the thermometer so that the bulb and the substance to be melted will be side by side. Place a small glass beaker nearly full of water on the sand-bath over the lamp. As the water approaches the melting-point of the substance, stir it gently with the thermometer, watching closely for any change in the appearance of the contents of the tube. On melting, it loses its opacity; and, just as soon as that change is observed, the reading of the thermometer must be taken. Now let the water cool, and record the temperature at which opacity returns. Repeat the experiment several times, and use the average of the results.

Suitable capillary tubes can be easily made by drawing out, in a flame, a piece of soft glass tubing till the diameter is about one millimetre.

254. Exercise. — Determine the boiling-point of a liquid, as ether, naphtha, alcohol, turpentine, etc.

Fit to a test-tube of 25 mm. diameter a cork with two holes. In one of these openings, insert a thermometer, and, in the other, a glass tube bent at right angles to serve as an escape for the vapor of the liquid to be tested (Fig. 109). Fill the test-tube one-third full of the liquid, and insert the cork, with the thermometer-bulb two centimetres above the liquid. Apply heat, and, on vapor escaping freely from the bent tube, take the reading of the thermometer. If the vapor of the liquid is combustible, apply the heat by means of an oil or water bath.

Why not have the thermometer-bulb in the liquid? Try it, comparing results with those obtained by the above process. Repeat the experiment with the thermometer in the liquid, having first placed in the test-tube some fragments of some substance not acted on by the liquid. Inference.

For most liquids, the temperature increases 0°.0375 C. for an increase of one millimetre in the atmospheric pressure. Hence, to reduce the temperature of the boiling-point obtained above to that for a pressure of 760 mm., add to it $0°.0375(760 - b)$, in which b is the barometric pressure at the time when the observation was made on the temperature.

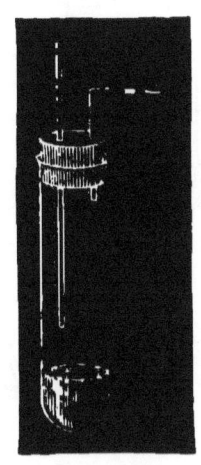

Fig. 109.

255. Exercise. — Determine the boiling-point of a saturated solution of some substance, as common salt, saltpetre, etc.

Proceed as in the last experiment. The thermometer-bulb must be in the liquid. Why? Pieces of broken glass, or some substance on which the liquid does not act, must be placed in the tube. Why?

256. Exercise. — Separate a mixture of two liquids having different boiling-points, as alcohol and water.

Set up an apparatus like that shown in Fig. 110. In the first flask, put the mixed liquids. This flask is connected to the second one by a bent tube of glass, and, in a similar manner, the second one is joined to the third. The first flask is directly over the lamp, the second one stands on a water-bath, and the third one stands in a vessel of cold water to keep it at a low temperature. The liquid in the first flask is made to boil gently, while that in the second one is kept at a temperature intermediate between the boiling-points of the two liquids, a condition easily determined by having a thermometer inserted through the cork into the flask.

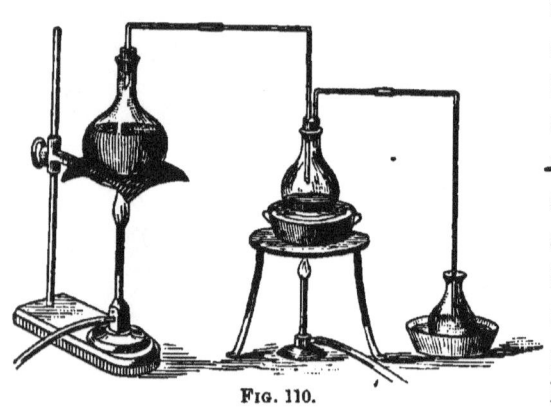

FIG. 110.

After a quantity of liquid has been collected in the third flask, examine it to ascertain whether the experiment has been successful.

A very convenient form of condenser is shown in Fig. 111, known as Liebig's. It consists of a large tube kept filled with cold water flowing from some convenient reservoir, through which passes the delivery-tube from the flask containing the liquids to be separated. The cold water enters the enveloping-tube at the lower end, and escapes from the upper end.

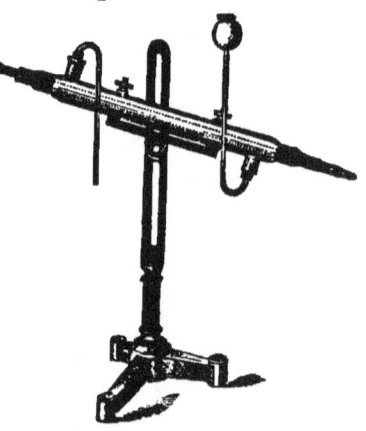

FIG. 111.

HEAT. 153

VII. RADIANT HEAT.

257. Apparatus. — Air-Thermometer, Leslie's Cubes, Florence Flasks, Differential Thermometer, etc.

258. Exercise. — Determine how the temperature of a substance affects its power to radiate heat.

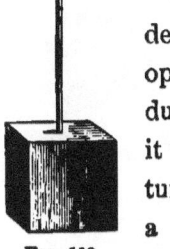

Fig. 112.

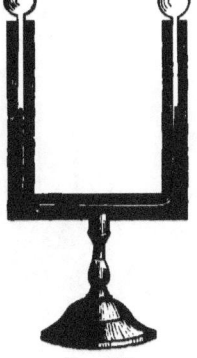

Fig. 113.

Construct of tin or copper a cubical box one decimetre on each edge, with an opening in one side for the introduction of water (Fig. 112). Fill it with water, bring its temperature to 10° C., and place it at a distance of, say, 3 cm. from a Differential, or Air, Thermometer (Fig. 113), observing the effect. Raise the temperature to 20° C., and again read the thermometer. Similarly test 30° C., 40° C., etc. Compare the indications of the air-thermometer with the readings of the mercurial one in the tank to see if there is any law connecting the two.

All air-currents must be carefully excluded.

259. Exercise. — Ascertain how distance affects the intensity of radiant heat.

Construct of tin or copper a tank 5 cm. thick by 50 cm. square. Fill this with boiling water, and place it at a distance of a few centimetres from the face of a **Thermopile** (Fig. 114), having a conical mouth-piece to concentrate the heat rays on the face. Observe the deflection of the galvanometer connected with the thermopile. Test the effect of increasing the distance of the tank from the thermopile. Is the surface which sends its heat to the pile the same for all distances of the tank? How does increasing the distance of the pile from the tank affect the area of the surface radiating its heat to the pile? What law for intensity of radiant heat can be inferred from the results?

The greatest distance that the tank should be placed from the pile is that at which the sides of the conical reflector,

FIG. 114.

on being extended, just touch the edges of the face of the tank. A sensitive air-thermometer can be substituted for the thermopile by adapting to it a funnel-shaped reflector such as accompanies the pile, supporting it in a burette-holder.

260. Exercise. — Employing the apparatus of the last experiment, place a large sheet of cardboard between the tank of water and the thermometer, and record the effect on the latter. What inference as to direction of radiation can be made?

HEAT. 155

261. Exercise. — Ascertain if air is essential to the transmission of radiant heat.

Fit a cork with two holes to a large Florence flask. Insert a thermometer in one of these openings so that the bulb is near the centre of the flask. In the other opening, fit a glass tube, connecting it by rubber tubing to a good air-pump. Exhaust the air as completely as possible, record the reading of the thermometer, then plunge the flask into hot water, and again read the thermometer. Ascertain how much the thermometer would be affected if the air were not exhausted from the flask. Inference.

262. Exercise. — Determine the law for the reflection of heat.

Place a tank like the one employed in Art. 258, filled with hot water, on a proper support, adjusting the height to that

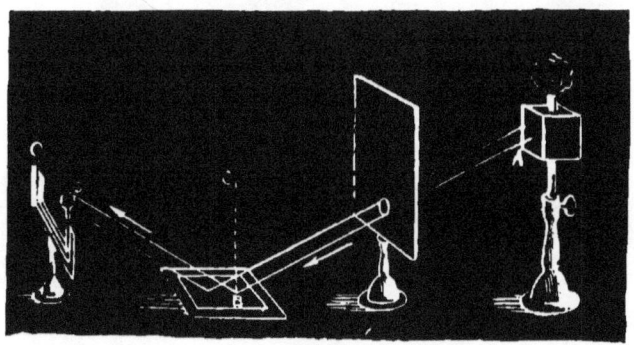

FIG. 115.

of the bulb of a differential thermometer. Half-way between the thermometer and the face of the tank, place a mirror horizontally. Between the mirror and the tank, support a sheet of cardboard in which a hole 3 cm. in diameter has been cut at the point where a line joining the centre of the face of the cube and the centre of the mirror intersects it (Fig. 115). What must be the course of the heat rays which reach the thermometer? Ascertain if moving either the radiator or the thermometer affects the readings of the thermometer. What law expresses the facts observed?

Ascertain if different kinds of reflectors are equally good.

Test the effect of fineness of polish.

Test the effect of increasing the size of the incident angle.

263. Exercise. — Compare the absorptive powers of different substances.

Coat with lamp-black one of the bulbs of a differential thermometer by mixing the lamp-black with thin shellac varnish.

FIG. 116.

Place the blackened bulb in the focus of a concave reflector such as is frequently placed behind wall-lamps (Fig. 116). At a short distance in front of the reflector, place the cubical tank of Art. 258, filled with hot water. Compare the effects of the heat on the plain bulb and the blackened one.

Try, in succession, the following substances as coverings for one of the bulbs: India-ink, tin-foil, foil of other metals, etc.

Arrange the substances tried, in the order of their absorptive power as shown by these experiments.

A differential thermometer, which will serve for these experiments, may be made as follows : —

Take two small flasks, and join them by a stout glass tube about 30 cm. long and 1 mm. bore, with its ends bent at right angles. Seal the flasks with wax to make them air-tight after introducing within the tube two or three drops of colored water. Attach a scale of equal parts to the horizontal portion of the connecting-tube, and mount the apparatus on a support.

A simple way of comparing the absorbing power of substances is the following : —

Cut two pieces of bright tin plate 10 cm. square. Saw in a narrow board two slits 10 cm. apart, and mount the plates in them, having first coated the inner face of one of them with the substance to be tested. Stick with wax, using as little as possible, balls of equal size on the outside of each at about the centre of the face. Now place a hot iron, as a soldering-iron, midway between the two plates. The better absorbent will be indicated by the melting of the wax holding the ball.

264. Exercise. — Fill the chemical tank (Fig. 234) with alum-water, place it between the sun and a large reading-glass, and focus the rays on a sheet of paper. Compare the effect with that obtained when the alum-water is omitted. Try the effect of a solution of iodine in carbon disulphide.

265. Exercise. — Ascertain if the radiating surface affects the rate of cooling.

Take two tanks similar to that of Art. 258, and blacken the faces of one with lamp-black, leaving the other bright. Fill each with water at, say, 80° C. ; insert in each a thermometer, and record the readings every five minutes for half an hour. Inference.

Two bottles of the same shape and size may be used as substitutes.

266. Exercise. — Compare the radiating power of substances.

Using the apparatus of Art. 265, coat one face of the tank with lamp-black, a second with white lead, a third with white paper, and the fourth with cotton cloth. Fill the tank with boiling water, and compare the radiating power of the four faces by observing the effects produced on the thermometer as each face is turned successively toward it for five minutes. The initial temperature of the water in the tank must be the same for each face.

Four equal bottles coated with the above substances respectively may be used as substitutes.

A radiometer may be substituted for the thermometer in many of these experiments in radiant heat, the intensity of radiation being indicated by the speed with which the vanes rotate.

267. Exercise. — Determine if the rate of cooling of a body is affected by the amount by which its temperature exceeds that of the surrounding air.

Take a cylindrical tin can of about one litre capacity, such as is used in preserving fruit, remove one end by unsoldering it, and fit to it a wooden cover, through the centre of which is a small hole of sufficient size to admit the bulb of a thermometer (Fig. 117). Blacken thoroughly the inside of the can, as well as the under side of the cover, by holding it in the smoke from burning gum-camphor. Select a thermometer having a large bulb as the object whose rate of cooling is to be studied. Slide over its stem a cork selected to fit the hole in the cover of the can, that the thermometer may be supported with its bulb at about the centre of the vessel. The tin can is designed to exclude air-currents from the cooling body. To maintain the surrounding air at a practically constant temperature, sink the vessel nearly to its top in a large vessel of water. Take the temperature, which will be that of the chamber, then introduce within the chamber the thermometer heated to, say, 80° C., by dipping it into hot water, carefully drying it with a cloth before putting it in place. Record the reading of the thermometer for every half-minute. In a parallel column, write the differences between these readings and the temperature of the water. These will

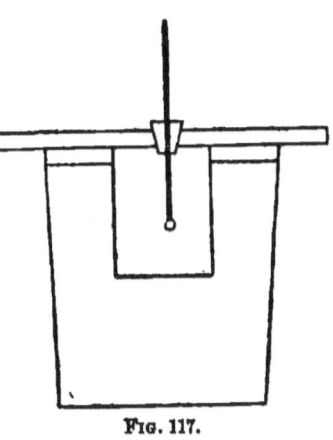

Fig. 117.

be the excess of the temperature of the object over that of the surrounding air for each half-minute. In a third parallel column, write the number of degrees the temperature of the object has fallen each half-minute. Finally, ascertain the ratio of the excess of the temperature of the cooling body above the temperature of the chamber, for each half-minute, to the fall of temperature during each corresponding half-minute, and place the results in a fourth parallel column. A comparison of these ratios will enable you to decide whether the rate of cooling is in any way connected with the temperature of the surrounding medium.

160   PRACTICAL PHYSICS.

VIII. CALORIMETRY.

268. Apparatus. — Beakers, Thermometer, Florence Flasks, Metal Balls of different substances and of the same diameter, etc.

269. Exercise. — Show that the capacity of water for heat is nearly constant; that is, the heat required to raise a given quantity of water a given number of degrees in one part of the thermometric scale is able to raise the same quantity nearly the same number of degrees in any other part of the scale.

Balance two thin glass beakers of about one litre capacity each, on the opposite pans of a balance, and pour into each 400 grammes of water. Let one be brought to the temperature of the room, and the other heated to about 60° C. Replace the beakers on the balance, and adjust the weights which have been disturbed by evaporation. Now take the temperature of each, observing the heated one last. Why? Then, without delay, pour the heated one into the other, stirring constantly with the thermometer for a few seconds, and record the temperature of the mixture. The difference between the temperatures of the mixture and the hot water will be the number of degrees lost by the hot water. Likewise the difference between the temperatures of the mixture and the cold water will be the number of degrees gained by the cold water. The average of the initial temperatures will nearly equal that of the mixture; the discrepancy being due to the heat absorbed by the cold beaker, that lost by radiation, and possibly that the amount of heat required to raise cold water one degree is different from that required to produce the same change in hot water.

Repeat the experiment, with the change that the cold water is poured into the hot, to ascertain how much the heat in the warmer beaker influences the result. Then the heat in the glass will go to raise the temperature of the cold water, and hence increase the temperature of the mixture.

In subsequent experiments, methods of correcting for absorption and radiation are given. These corrections carefully applied to this problem have shown that the capacity of water for heat is not strictly constant, but increases slightly with the temperature.

270. Exercise. — Mix unequal quantities of water at different temperatures, and ascertain the temperature of the mixture. Compare the result with that obtained by computation on the supposition that all the heat lost by the hot water went to the cold water.

To compute the temperature, divide the total number of heat units in the two quantities of water by the total quantity of water.

271. Exercise. — Determine the water-equivalent of a vessel; that is, find how much water will equal it in capacity for heat.

Dry the vessel thoroughly, and let it come to the temperature of the room, which will be indicated when a thermometer suspended in it registers the same as when without it. Then pour in it a known quantity of water, as .4 kg., at a known temperature, say, 40° C., and record the temperature after the lapse of about thirty seconds. The fall in temperature is due to heat having been absorbed by the vessel, and to radiation, while pouring in the water and stirring. This may be allowed for as follows: Before pouring the water into the vessel, ascertain the fall of temperature in one minute, then observe the fall in one minute following the period of thirty seconds allowed for the heat to be communicated to the vessel. Then the radiation rate may be taken as the average of these two losses. One-half of this will be the loss from radiation during the thirty seconds, and will be the amount to be added to the observed temperature of the water in the vessel to give the correct temperature for the water, and hence of the vessel. The number of calories of heat consumed by the vessel will be the loss of temperature on the part of the water, times the amount of water. This divided by the gain of the vessel will give the number of calories required to change the temperature of the vessel one degree, and hence will be the number of kilogrammes of water to which the vessel is equivalent.

272. Exercise. — Determine the thermal capacity of a substance, as lead, copper, etc.

Make a loose coil of a known weight of the metal in sheet form, and suspend it, by a thread, in boiling water for a sufficient time to acquire the temperature of the water. The temperature of the boiling water must be observed. Why? Pour into a beaker, whose water-equivalent has been determined as in Art. 271, a known weight of water of the temperature of the room, and of sufficient amount to cover the coil entirely when placed within it. Now transfer to it the heated coil, stir the water around thoroughly with a thermometer, and record the reading when it ceases to rise, and also the time occupied in thus equalizing the temperatures. Keep up the stirring for one minute longer, that the effect of radiation may be observed. Half of this loss multiplied by the time occupied in securing equalization must be added to the observed temperature, to give the correct temperature produced in the water by the heated coil. Now add to the weight of water used, the water-equivalent of the beaker, and multiply this by the gain in temperature to obtain the number of calories of heat imparted by the coil to the water and beaker. This product divided by the number of degrees lost by the coil will give its capacity for heat.

273. Exercise. — Compare, by Tyndall's method, the thermal capacities of several metals.

Make a ball out of each of the metals lead, tin, zinc, bismuth, antimony, etc., by the aid of a pair of large bullet-moulds, and an iron spoon in which to melt the metal. Reduce them to the same weights by cutting off a part of the ball. Solder a fine wire handle to each. Mould, in a shallow pan, a cake of wax about 5 mm. thick. Support the cake in a horizontal position by its edges. Heat the balls in boiling water till they have acquired the temperature of the water. Place them simultaneously on the wax, several centimetres apart, and observe the time occupied by each in melting through the cake (Fig. 118). Some of the balls may not get through; in that case, the depth the ball melts into the cake must be noted. The drop of water adhering to each ball must be removed before placing the ball on the wax cake. The handles should be attached at the point where the weight-adjustment was made, so that like surfaces, in every case, will be in contact with the wax. Only approximate results can be secured by this method.

FIG. 118.

274. Exercise. — Determine the specific heat of a liquid.

First Method. — Make a loose coil of some metal, as lead, whose specific heat is known, and then, proceeding as in

Art. 272, find its thermal capacity with reference to the liquid in question by substituting the liquid for the water. The quotient of the heat capacity when compared with water, by that when compared with the liquid, will give the thermal capacity of the liquid with reference to water.

Second Method. — Procure two test-tubes of the same size; in one of them pour a few grammes of water, and, in the other, an equal volume of the liquid under examination. In each test-tube, insert a thermometer; then place the tubes in a vessel of water, at, say, 70° C., as indicated by a third thermometer, and note the time required for the contents of the tubes to acquire that temperature. As these substances differ in density, it will be necessary to reduce these times to those for equal weights by dividing each by the density of the corresponding liquid. The ratio of these times will be the specific heat of the liquid. Compare the results given by this method, with those given by the first.

Try turpentine, ether, kerosene, benzine, glycerine, etc.

275. Exercise. — Determine the latent heat of water; that is, find the number of calories of heat disappearing during the melting of one kilogramme of ice.

Pour into a beaker a known quantity of water, say 500 grammes, and raise its temperature to, say, 70° C. Set the beaker on a board, and observe the fall in temperature during one minute. Now add ice in small pieces till about 200 grammes have been introduced, constantly stirring the water with the thermometer. Each piece of ice should be wiped with a dry cloth before putting it in the beaker, to avoid introducing any water in the liquid form. The time occupied in putting in the ice should be as short as possible. As soon as the ice is melted, take the temperature of the water, and record the time

HEAT. 165

occupied. Also obtain the fall due to radiation during the next minute. The amount of ice introduced is the increase in weight of the beaker and contents. The following example will show the method of making the computation: —

Amount of water in beaker	500 grammes.
Temperature of water	75° C.
Fall in temperature during one minute	$1\frac{1}{2}$° C.
Temperature of water taken as soon as ice melted	32° C.
Fall in temperature during next minute	$\frac{1}{2}$° C.
Time required to melt ice	3 minutes.
Amount of ice introduced	179 grammes.
Water-equivalent of beaker previously determined	34 grammes.

Data corrected for radiation and absorption: —

Amount of water, $500 + 34 = 534$ grammes.

Temperature of water before introducing ice, $73\frac{1}{2}$° C.

Temperature of water after introducing ice, $32 + 3\left(\frac{1\frac{1}{2} + \frac{1}{2}}{2}\right) = 35$° C.

$.534(73\frac{1}{2} - 35) = 20.559$ calories of heat consumed;

$.179 \times 35 = 6.265$ calories of heat consumed in raising the ice from 0° to 35°;

∴ $20.559 - 6.265 = 14.294$ calories of heat consumed in melting .179 kg. of ice,

∴ $14.294 \div .179 = 79.9$ calories of heat consumed in melting 1 kg. of ice.

276. Exercise. — Determine the latent heat of steam.

Fit to a Florence flask of about one litre capacity a delivery-tube of the form shown in Fig. 119. Fill the flask half full of water, and support it on the iron stand. The delivery-tube should reach nearly to the bottom of a beaker containing a known quantity of water, say 400 grammes, at the temperature of the room. Apply heat to the flask, and, as soon as a strong jet of steam issues from the delivery-tube, let it enter the water in the beaker. Stir the water constantly and gently with the thermometer; and, when the temperature has risen a certain number of degrees, as 20° C., record the time occupied, and also the fall from radiation during the next minute. Then ascertain the increase in weight, and also the temperature of steam. The following example will show the method of making the computation: —

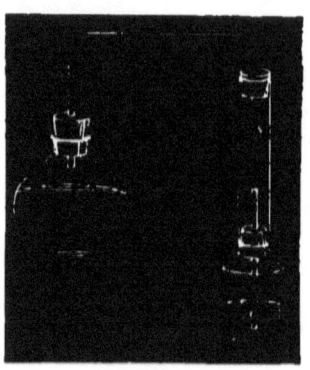

FIG. 119.

Amount of water in beaker . . . . 400 grammes.
Water-equivalent of beaker previously determined. 34 grammes.
Temperature of water in beaker . . . . 21°.75 C.
Temperature of water after steam ceases to enter . 41°.75 C.
Loss in one minute from radiation . . . . . . 0°.25 C.
Time occupied in admitting steam . . . . . . 4.25 minutes.
Amount of steam introduced . . . . 14 grammes.
Temperature of steam . . . . 99°.25 C.

Temperature of water in beaker corrected for radiation,

$$41.75 + \frac{.25}{2} \times 4.25 = 42°.28 \text{ C.}$$

Amount of water corrected for beaker-equivalent, 434 grammes.

.434 × 20.53 = 8.91 calories of heat imparted to the water.

.014(99.25 − 42.28) = .8 calories of heat derived from the water produced by the steam in condensing and cooling from 99°.25 C. to 42.28 C.

∴ 8.91 − .8 = 8.11 calories of heat derived from .014 kg. of steam condensing to water at the boiling-point.

Hence

8.11 ÷ .014 = 579.3 calories of heat latent in 1 kg. of steam.

It will be observed that the large piece of tubing inserted in the delivery-tube will retain the water produced by steam condensing before reaching the water in the beaker. To admit this would introduce serious error. Why?

IX. ARTIFICIAL COLD.

277. Exercise. — Dissolve in three parts, by weight, of water not warmer than 10° C., a mixture of two parts of pulverized ammonium nitrate and one part of ammonium chloride. Stir the mixture with a test-tube containing water, observing the reading of a thermometer placed in the tube. Account for the low temperature.

278. Exercise. — Dissolve common salt in water, and see if it affects the temperature of the water in any way. Try potassium nitrate, ammonium nitrate, etc. Try some substance not having a crystalline character, as dextrine, gum-arabic, etc. Account for the difference in the results.

279. Exercise. — Tie a piece of cotton-wool about the bulb of a mercurial thermometer. Wet the wool with ether, and note the effect. Explain.

280. Exercise. — Pour water at the temperature of the room into a porous cup, such as is used in the construction of galvanic batteries. A new unglazed flower-jar will answer the purpose if the drainage-hole is closed up. Take the temperature every five minutes for half an hour, and also that of the room at the same time. Account for the difference.

281. Exercise. — Place a bell-jar on the table of the air-pump, observe closely the degree of transparency, then exhaust the air, watching for any change in the appearance of the jar. Now admit the air, introduce a thermometer, take the temperature, and, after exhausting the air, again take the temperature. Why is it lower? Account for the phenomenon seen during the first exhaustion. What applications can you find in the theory of clouds, of the truth brought out in this experiment?

282. Exercise. — As in Art. 278, dissolve sodium sulphate in water, determining the effect on the temperature. Now prepare a saturated solution of sodium sulphate by dissolving a few grammes of the substance in an equal weight of hot water. Set the solution aside in a vessel to cool, having first poured a thin layer of oil over the top. If not disturbed, it will cool to the temperature of the room without crystallizing. After waiting long enough to be certain that the solution has reached the temperature of the room, introduce a thermometer, and observe the effect. Explain.

Try sodium acetate in the same manner.

283. Exercise. — Test some of the following freezing mixtures : —

1st, Common salt, by weight, one part; snow or pounded ice, two parts. The temperature will sink to $-20°$ C.

2d, Common salt, by weight, five parts; ammonium nitrate, five parts; snow or pounded ice, twelve parts. The temperature will sink to $-31°$ C.

HEAT.

3d, Crystallized calcium chloride, by weight, two parts; snow or pounded ice, two parts. The temperature will sink to $-40°$ C.

4th, Sodium phosphate, by weight, nine parts; dilute nitric acid, four parts. The temperature will sink to $-29°$ C.

5th, Sodium sulphate, by weight, six parts; ammonium nitrate, five parts; dilute nitric acid, four parts. The temperature will sink to $-26°$ C.

In these mixtures, the substances are supposed to be at the temperature of $10°$ C. If, however, they are previously cooled down, a lower temperature still can be obtained in most cases.

CHAPTER V.

MAGNETISM AND ELECTRICITY.

I. MAGNETS. — POLARITY. — INDUCTION.

284. Apparatus. — Bar-Magnets, Needles, Thin Plates of Different Substances, Short Rod of Soft Iron, Carpet-Tacks, etc.

285. Exercise. — Touch one end of a bar-magnet to a pile of carpet-tacks, and record the effect. Try the other end of the magnet. Count the number of tacks adhering in each case, and obtain the average of a number of trials. Inference. Ascertain if any piece of steel will affect the tacks in the same way.

Make a paper stirrup, place the bar-magnet in it, and suspend the bar-magnet by a thread, thus giving the magnet freedom of motion. Now bring near each end, in succession, an iron nail. Repeat the experiment, the nail and magnet having changed places. Does the magnet attract the nail, or does the nail attract the magnet?

Magnets for the purpose of this experiment, as well as the following ones, may be made by stroking, from end to end, and always in the same direction, large steel nails, or short steel rods, with one pole of an electro-magnet (Art. 382).

286. Exercise. — Float a common darning-needle on a piece of cork in a glass vessel of water, and ascertain, by making several trials, if, after coming to rest, it points in any one way in preference to another. Now stroke the needle, from end to end, with one pole of a magnet, and repeat the tests. Inference.

287. Exercise.— Magnetize a large sewing-needle by stroking it a few times with one pole of a magnet. Pull a silk fibre several centimetres long out of silk floss, by untwisting it, and attach it, by a very little wax, to the needle, so that, on holding it up, the needle will take a horizontal position. Cement the other end of the fibre to one end of a small glass tube passing through the centre of a cork. Fit this cork to the top of a large lamp-chimney or bottle. The length of the fibre should be such as to bring the needle within half a centimetre of the base. This instrument will serve as a magnetoscope. A pocket-compass, or a magnetic needle, mounted as in Fig. 120, may be used instead.

FIG. 120.

Magnetize a large darning-needle by stroking the half toward the point with the south-seeking pole of a magnet, and the other half with the north-seeking pole. Ascertain which is the north-seeking pole of the needle as in the last experiment. Now bring its north-seeking pole near that pole of the magnetoscope, and record the effect. Try the other pole. What law seems to govern the action of magnets toward each other?

288. Exercise. — Hold the north-seeking pole of a strong magnet near the north-seeking pole of a feebly magnetized needle, and ascertain if the effect is in harmony with the law of magnetic action. Now re-examine the polarity of the small needle. Inference.

289. Exercise. — Ascertain if a strong permanent magnet will attract iron-filings or tacks through thin plates of mica, wood, paper, glass, copper, zinc, iron, etc.

290. Exercise. — Bend a strip of heavy sheet-iron, 15 cm. by 5 cm., into a ring. Suspend, by a thread, from a wooden support, a small piece of iron, and adjust it to hang at the centre of this hollow cylinder as it rests on the table. Now hold, opposite the ball of iron on the outside of this cylinder, a strong magnet, and compare the effect on the ball with that produced if the ring is removed and the magnet is held at the same distance as before. Try a brass or zinc ring, and see if the effect is the same.

291. Exercise. — Hang as many tacks as possible from the pole of a magnet supported in a horizontal position above the table, using a clamp for the purpose. Hold up to it a sheet of paper carrying carpet-tacks, and notice how many adhere. Now place a second magnet beneath this one, so that its opposite pole is exactly under the one supporting the tacks, and ascertain if the power of the first magnet to support tacks has been affected. Now place the magnet so that like poles are opposite, and repeat the tests. Explain.

292. Exercise. — Hold one end of a rod of soft iron near one pole of a magnet, and, while in that position, dip the other end of the rod into iron-filings. Record the result. Test the soft-iron bar for polarity while in this position, using a pocket-compass for the purpose. Note the effect of removing the magnet.

Vary the experiment by holding the magnet near to the middle of the rod, and testing the whole rod with iron-filings. Also test for polarity.

MAGNETISM AND ELECTRICITY. 173

Arrange two such rods end to end along on a table, and not quite in contact. Near one end of the row, hold a strong magnet. Test the last one of the row with iron-filings. If placed with its end projecting over the edge of the table, a dish of filings can be held to it without trouble. Also test each end of these bars with a compass-needle to ascertain their polarity.

II. NATURE OF MAGNETISM.

293. Apparatus. — Steel Bar, Magnet, Needles, Steel-Filings, etc.

294. Exercise. — Ascertain if it sensibly affects the weight of a bar of steel to magnetize it.

295. Exercise. — Take a piece of No. 16 iron wire about 30 cm. long, anneal it carefully by heating it red-hot, and letting it cool very slowly. Turn at right angles one centimetre at each end for convenience in holding. Now stroke it carefully with a magnet several times, and test its power to lift iron filings. Then give the wire a sudden twist, and again test it.

Magnetize a knitting-needle; test its power to pick up small iron tacks; then, holding one end firmly, make the needle vibrate by plucking the free end, and again test its power to lift small iron tacks.

296. Exercise. — Magnetize a needle, or a piece of steel wire, and dip one end into a dish of small iron tacks, noticing the number adhering as you lift it out. With a pair of crucible tongs, hold the needle in a flame till red-hot, and then plunge it quickly into the tacks, and notice if it picks up as many as before. Test again after the needle has become cold.

Vary the experiment by ascertaining if a magnet will attract a small nail when it is red-hot. Also vary the experiment by letting a heated steel needle cool, having placed it, when red-hot, in a line between the opposite poles of two magnets.

From the nature of heat, what does this experiment suggest regarding magnetism?

297. Exercise. — Magnetize a knitting-needle, and roll it in iron-filings, recording the result. Now break the needle into two equal pieces, and test them. Break these pieces, and test them. Why do not filings adhere at the centre? What view of the nature of magnetism is here supported?

298. Exercise. — Fill a glass tube 1 cm. in diameter and 10 cm. long with very short pieces of steel wire, or with steel filings, closing the ends with a cork. Stroke the tube with a powerful magnet, and test it for polarity by holding its ends in succession near one pole of the magnetoscope. Now shake up the contents of the tube thoroughly, and again test its polarity. Inference.

The coercive force can be imitated by mixing a little oil with the filings. It will be found more difficult to magnetize and demagnetize the tube.

III. THE MAGNETIC FIELD.

299. Apparatus. — Magnets, Short Magnetic Needle mounted, Magnetoscope, etc.

300. Exercise. — Map out the magnetic field of a magnet.

Place a short bar-magnet beneath a stiff sheet of writing-paper supported on two wooden bars as thick as the magnet,

MAGNETISM AND ELECTRICITY. 175

and sift iron-filings evenly over it from a thin muslin bag containing them, tapping the paper gently to facilitate the movements of the filings. Study the figures obtained in this way for a horse-shoe magnet, a disk-magnet, two bar-magnets placed parallel to each other, with like poles adjacent, and separated by about two centimetres; two magnets with unlike poles adjacent, horse-shoe magnet with its armature on, a bar-magnet with a bar of soft iron near one of its poles, an iron ring when a powerful magnet is near it, a bar-magnet held vertically, etc.

Permanent copies of these figures may be made as follows: —

Brush a sheet of printing-paper over with a solution of tannin, and place it carefully on the figure after removing the magnet. Place on this a sheet of heavy blotting-paper, and apply a slight pressure. On lifting off the paper, most of the filings will adhere to it, and can be brushed off when dry, leaving dark marks on the paper.

Place a small magnetic needle successively in different parts of the magnetic field, as marked out by the filings before removing the magnet, and notice its position with reference to the line of force passing through the support.

A suitable needle may be constructed as follows: Straighten a piece of watch-spring, cut from it a lozenge-shaped piece 15 mm. long, and drill a hole 1 mm. in diameter through the centre. Cement, with a little wax, a small cap exactly over this hole to serve as a bearing. To make such a cap, heat the end of a very small glass tube in a Bunsen flame till the glass, by contracting, closes the end. Cut off, with a file, a piece about 3 mm. long, and you will have a cap with a smooth conical hole in it for the reception of the supporting pivot. For a pivot, break off the point of a sewing-needle, and insert it in a piece of wood 1 cm. square and 2 mm. thick for a base.

If the needle is not level after magnetization, adjust it with a little wax.

301. Exercise. — Represent, by a curve, the change in the magnetic strength as you go from the poles of a magnet to the centre.

Magnetize, as uniformly as possible, a heavy knitting-needle. This is best done by the method known as *Divided Touch*. It consists in fixing the needle lengthwise between the opposite poles of two permanent magnets, and, while under their induction, stroke the needle, each half, with the pole of another magnet of the same name as the corresponding inducing pole, the stroking-magnets being held in the hands at an angle of about 30 degrees with the needle. The stroking must begin at the centre of the bar, the poles being lifted at the ends, and brought back in an arch to the centre. The stroking should be repeated on the opposite side of the needle.

A Coulomb's magnetoscope will be required for this experiment. To construct one, magnetize to saturation a small cylinder of glass-hard steel about 10 mm. long by 5 mm. in diameter. Suspend this, by a silk fibre, in a glass tube about 15 cm. long set into a wooden base. A wire hook sliding through a cork fitted to the tube will serve to support the fibre. The magnet is cemented in a wire stirrup, and should hang about midway between the ends of the tube. To prevent it from swinging, connect it, by a second fibre, to a small lead disk resting on the floor of the tube.

Place the magnetoscope on the table, and, by means of a magnet, produce a small displacement of the magnetoscope-needle in order to set it in vibration through a small arc. Ascertain the number of oscillations made in thirty seconds, while vibrating under the influence of the earth alone, averaging at least three trials. Now support the magnetized knitting-needle in a vertical position, bring its centre near the magnetoscope and in the magnetic meridian. As before, ascertain the number of vibrations made by the magnetoscope-needle in thirty seconds, being careful to set it vibrating through an arc of the same size as before.

In a similar way, test the ends of the magnet, and four or more places between the centre and the ends. Care must be taken not to vary the distance between the magnet and the magnetoscope, and not to change the position of the magnetoscope on the table. The bringing of like poles together must be avoided so that attraction, and not repulsion, will actuate the instrument. The position of the points on the magnet, which were tested, must be known relatively to the length of the magnet.

Now square the number of vibrations in each case, and subtract the square of the number representing the magnetism of the earth. The numbers thus obtained represent the relative magnetic intensities at these points, since the attractive force varies as the square of the number of vibrations. Draw a straight line 10 cm. long to represent the magnet. At points situated like those tested on the magnet, erect perpendiculars. Make the one at the plus pole 2 cm. long to represent the magnetic intensity at that point, and give the others lengths proportional to the numbers just obtained. Through the extremities of these lines, sketch a curve, and you have a graphic representation of the change in magnetic intensity as you go from the middle of a magnet toward either end.

302. Exercise. — Proceeding as in the last experiment, compare the strengths of the north-seeking poles of two bar-magnets.

303. Exercise. — Proceeding as in Art. 301, determine the number of vibrations made by the needle of the magnetoscope, in thirty seconds, when distant from the pole of a magnet 2 cm., 3 cm., 4 cm., etc., respectively. Compare the ratios of these numbers representing the strengths of the magnetic field at these distances from the pole with the ratios of the distances. If the experiment is carefully conducted, one set of ratios will be found to be the square of the other set taken in an inverse order. What law is supported by this fact?

304. Exercise. — Determine, from the law of inverse squares, the direction a magnetic needle will have when at rest in the neighborhood of a bar-magnet.

Magnetize a knitting-needle, and lay it on a straight line drawn on a sheet of paper, marking on this line the position of the poles N and S of the magnet at one-tenth of the length of the needle from the end (Fig. 121), as the poles of a magnet are not at its ends. Take A as a point in the magnetic field.

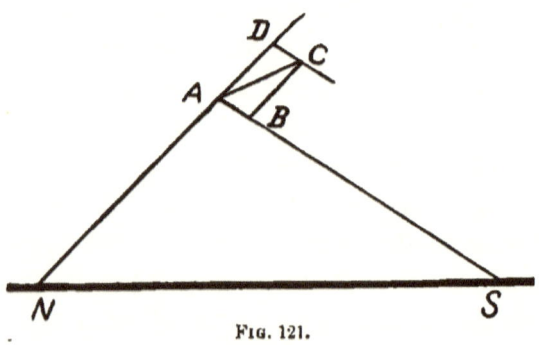

Fig. 121.

The north-seeking pole of a magnet at A would be attracted by S along the line AS, and repelled by N along AN. Measure AS and AN, and then lay off AB any convenient number of centimetres, and AD as many times greater than AB as AS squared is greater than AN squared. Complete the parallelogram ABCD, and the diagonal AC will represent, by its length and direction, the magnitude and the direction of the resultant magnetic force at A. Verify the conclusion by placing a short magnet at A free to turn, such as was employed in Art. 300.

Locate a number of these lines of direction, and compare them with the lines of force for the magnet as marked out by the use of iron-filings.

IV. TERRESTRIAL MAGNETISM.

305. Apparatus. — Dipping-Needle, Iron Bar, Compass, etc.

306. Exercise. — Determine the dip of the magnetic meridian.

Fasten, with a little wax, a thread to the middle of an unmagnetized knitting-needle, finding, by trial, a point of attachment in which the needle takes a horizontal position. Now stroke the needle with a magnet, and suspend it by the thread attached. It now acts as if one end has become heavier than the other. To obtain accurately the amount of this dip, a **Dipping-Needle** (Fig. 122) will be needed.

FIG. 122.

This, when placed in the magnetic meridian, shows, on a graduated arc, the amount of dip.

307. Exercise. — Procure a thoroughly annealed iron bar 75 cm. long, showing little or no polarity when tested with a magnetic needle while the bar is supported horizontally in an east-and-west line. Now support this bar in a position parallel to that taken by the dipping-needle of the last experiment, and again test it for polarity. Turn the bar end for end, and again test it for polarity. Inference.

V. FRICTIONAL ELECTRICITY.

308. Apparatus. — Glass Rods, Sealing-wax Rod, Silk, Flannel, Pith-Balls, Electroscope, etc.

309. Exercise. — Balance a common yard-stick, or a strip of lath, on the point of a needle projecting through a cork in a bottle. Hold, near one end of the stick, a rod of sealing-wax immediately after rubbing it briskly with flannel. A rubber comb or ebonite ruler may be substituted for the sealing-wax.

Try a glass tube after rubbing it briskly with a piece of silk. The tube should be about 50 cm. long and 2 cm. in diameter, with the ends rounded, by softening them in a gas-flame, to prevent cutting the fingers when experimenting with it. The tube and the silk should be warmed before using. Glass free from sodium is best for the purpose.

Describe the effects these objects have on the balanced strip.

See if there is any connection discoverable between the mechanical energy expended in rubbing the rod, and the intensity of the state developed in it.

310. Exercise. — Excite each of the various rods used in the last experiment, and in the way there described, and immediately bring it near to a pile of pith-balls or small bits of tissue-paper.

The pith-balls can be obtained from the common elder, or from the corn-stalk. They may have a diameter from 5 mm. to 1 cm. Cut the pith with a very sharp knife in order that each ball may have a smooth surface.

MAGNETISM AND ELECTRICITY. 181

311. Exercise. — Make a hoop out of paper, say, 20 cm. in diameter and 8 cm. wide. Place it upon a table, and hold a glass rod excited by rubbing with silk a little below one extremity of a horizontal diameter of the hoop. Do not let the hoop touch the rod, but keep it at a distance of one or two centimetres from its outer surface. Describe the effect on the hoop.

312. Exercise. — Determine the law of electrical action.

Rub a glass rod with silk, and suspend it in a paper stirrup (Fig. 123). In like manner, excite a second glass rod with as little delay as possible, and bring it near one end of the suspended one. Note the effect. Now try two rods of sealing-wax or ebonite in the same way, and observe the effect. Finally, try one of glass, and one of sealing-wax, and observe their action toward each other. If we represent the electrification of glass by $+E$, and that of sealing-wax by $-E$, what law expresses the conclusions reached?

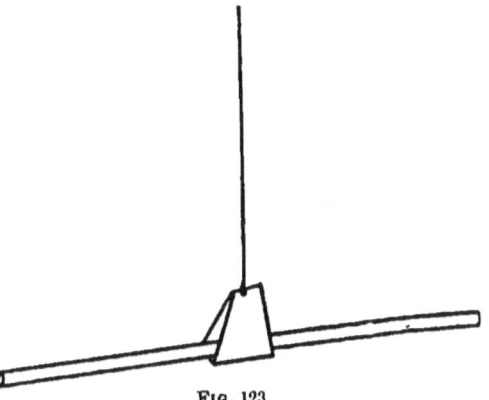

FIG. 123.

313. Exercise. — Draw a silk ribbon between two layers of warm flannel, and hold it near a wall or some object. Try two strips of ribbon hanging side by side. Explain.

314. Exercise. — Place a sheet of brown paper on a hot dry board. Brush it briskly, for a few moments, with a hair clothes-brush. Cut part of the paper into narrow strips free at one end, with a sharp knife, without touching the surface of that part with the hand. Observe the action of these strips toward one another. Explain.

315. Exercise. — Test various articles found in the room, for electrification, after rubbing them with either silk, fur, or flannel.

Select a Florence flask of about one litre capacity, thoroughly clean and dry it; fit to it a cork, through which passes a stout brass wire, on the outer end of which is a metal ball about 15 mm. in diameter, or a thick brass disk of the same diameter, having the edges neatly rounded off with a fine file (Fig. 124). File the end of the wire within the flask to a wedge form, the edge having a thickness of not over one millimetre. Cut, out of thin tin or aluminum foil, two strips, each 5 mm. wide and 6 cm. long, and paste them on the faces of the lower end of the wire so that they will hang side by side. Tin-foil (No. 4), such as is used by dentists, makes excellent strips for this purpose. Such an instrument is called an Electroscope. To use it, bring the article suspected of being electrified near the ball of the instrument; the divergence of the leaves shows electrification.

Fig. 124.

If a little strong sulphuric acid is put in the flask, it will tend to dry the air within it, and increase the duration of the charge.

316. Exercise. — Determine the kind of electrification of a body.

Charge an electroscope with the unknown kind of electrification, using a proof-plane for the purpose. A **Proof-Plane** (Fig. 125) is usually a small metal disk cemented to the end of a glass or ebonite rod. A serviceable instrument can be made by cementing a cent, after filing the edges smooth, to a glass tube about 15 cm. long. To charge the electroscope, slide the metal disk of the proof-plane along the surface to be tested, and then bring it in contact with the electroscope-knob. Repeat the operation till the ends of the leaves are about 1 cm. apart. Now excite strongly, by friction with silk, a dry glass rod. If the silk rubber is coated with amalgam (see Appendix, p. 360), a higher electrification can be secured. Charge the proof-plane by putting it in contact with the rod, and then, without delay, place this charged proof-plane in contact with the electroscope-knob. If the leaves diverge farther, the object in question was positively charged; if the leaves collapse, it was probably negative. As increased divergence is a more reliable indication of electrification than a decreased divergence, in case of the latter occurring it is best to repeat the test, using a rod of sealing-wax and flannel in place of the glass rod and the silk.

Fig. 125.

It is important that the electroscope be not charged too highly, as, in that case, a decreased divergence might mean that the intensity of the known charge in the proof-plane was less than that of the electroscope, and of the same kind, thus introducing doubt. Always charge the proof-plane as intensely as possible, in order that ambiguity may not arise.

Test the kind of electrification of the silk ribbon in Art. 313, the paper in Art. 314, common rubber excited by flannel, glass excited by flannel, stick-sulphur excited by flannel, sulphur excited by silk, rubber excited by silk, sealing-wax excited by cotton, paper excited by flannel, etc.

317. Exercise. — Excite a glass rod by rubbing it with silk, keeping the silk rubber from touching the hand by means of a square piece of sheet-rubber such as dentists use. Test the silk for electrification. Similarly test the flannel employed in rubbing sealing-wax, fur in rubbing any resinous substance, etc. Inference.

318. Exercise. — Compare the conducting power of substances for electricity.

Place one end of a thread of the substance whose conductivity is to be tested, in contact with the electroscope-knob, and the other in contact with a smooth metallic button supported on a rod of sealing-wax. A thread 50 cm. long may be used when possible, and, in case of some substances, it would be well to make it still longer. Now touch the button with a highly charged proof-plane, and decide, from the action of the electroscope, whether the substance is a good conductor, a semi-conductor, or a non-conductor.

Try silk thread, cotton thread, linen thread, woollen thread, glass thread, rubber, wood, paper, wire, graphite as found in a lead-pencil, any kind of thread wet with water, etc.

VI. STATICAL INDUCTION.

319. Apparatus. — **Insulated Metallic Cylinders and Balls, Electroscope, Metallic Cylindrical Pail,** etc.

320. Exercise. — Place end to end, and in contact, two metallic cylinders about 20 cm. long and 4 cm. in diameter, having rounded ends, each cylinder being supported on a varnished glass rod inserted in a wooden base (Fig. 126). Wooden cylinders smoothly covered with tin-foil make a cheap and efficient substitute, the foil being fastened on with flour-paste. Metallic door-knobs mounted on glass rods work well. As you hold near one end, that is, 2 or 3 cm. removed, a highly electrified glass rod, by means of a proof-plane and electroscope test the outer ends, and the point of contact, of the two cylinders for electrification. If found to be electrified, ascertain the kind. Now recharge the glass rod; and, as you hold it near the end of one of the cylinders, separate the cylinders by several centimetres. Test the ends of the separated cylinders for electrification, comparing results with those obtained when the cylinders were in contact. Explain.

FIG. 126.

This experiment will succeed only in a very dry atmosphere.

321. Exercise. — Connect one end of the insulated metallic cylinder of the last experiment with an electroscope by means of a wire, the ends of which are rounded, or bent around into a loop and soldered. Hold an electrified glass rod near the other end of the cylinder, and record the effect on the electroscope. Before removing the excited rod, remove the wire, handling it with some non-conductor. If the wire rests loosely on the cylinder and on the ball of the electroscope, it can be lifted off with a glass rod. Ascertain the electrical condition of the electroscope as well as of the cylinder.

322. Exercise. — Hold one finger on the ball of an electroscope, and bring near it an electrified glass rod. Remove the finger before taking away the rod. Determine the electrical condition of the electroscope. Try a rod of sealing-wax excited by flannel, in place of the glass rod.

This is a very desirable way of charging an electroscope with a known kind of electricity. The intensity of the charge can be varied by changing the distance the excited body is held from the ball of the instrument.

323. Exercise. — Determine how the amount of electricity induced on neighboring conductors compares with that of the inducing body.

Procure a metallic cylindrical pail free from points, and support it on a plate of well-dried glass to insulate it. This vessel should have a depth of about 25 cm. and a diameter of about 15 cm. It may be made of tin or copper. Connect the outside, by a wire, to the ball of an electroscope. Suspend by a silk thread a metallic ball, say, 5 cm. in diameter, having a smooth surface and positively charged, within the pail. Notice the indication of the electroscope. With a proof-plane and a second electroscope, determine the electrical condition of the two surfaces of the pail. Ascertain if the position of the electrified ball within the pail in any way affects the attached electroscope. Touch the ball to the bottom of the pail, and observe if the electroscope is affected. Now remove the ball, and test it for electrification. What must have been the amount of the electricity on the inside of the pail compared with that on the ball? What must have been the amount of electricity on the outside of the pail?

VII. ELECTRICAL DISTRIBUTION.

324. Apparatus. — Metal Beaker, Pane of Glass, Wide Brass Ring, Electroscope, Brass Chain, Insulated Metallic-covered Ball, Cylinder, and Cone; Proof-Plane, Insulated Metal Disks, etc.

325. Exercise. — Support on a dry pane of glass a metallic vessel of about one litre capacity, free from sharp edges. Charge it with electricity by rubbing an excited glass rod over it. Employing a proof-plane and an electroscope, test the two surfaces for electricity. Inference.

326. Exercise. — Bend into a ring a heavy strip of sheet brass 25 cm. long and 8 cm. wide, soldering the ends together, and filing the edges round. Suspend from the inside surface a pair of pith-balls, and a second pair from the outside surface. The inside pair of balls should be at the centre of the ring. Suspend the apparatus by a silk cord from a support. Now electrify, by friction, a glass rod, and rub it across the cylinder, observing the effect on the pith-balls. Inference.

Fig. 127 shows such a cylinder mounted on an insulated stand.

FIG. 127.

327. Exercise. — Support the metal vessel of Art. 325 on a dry pane of glass, and connect its outer surface with the ball of the electroscope by means of a wire. Suspend within the vessel, from a glass rod, a metre or less of brass chain. Charge the vessel with electricity till the leaves of the electroscope diverge widely; then lift up the chain so that less of it lies piled up on the bottom of the dish, observing the effect on the electroscope. Keeping in mind that lifting up the chain practically increases the surface of the vessel without decreasing the quantity of electricity, what inference can be drawn regarding the effect on potential produced by changing the amount of surface?

328. Exercise. — Determine whether the shape of a conductor in any manner affects the distribution of the electrical charge.

Cover smoothly with tin-foil a wooden ball 8 cm. in diameter, a conical-shaped piece 15 cm. long, with hemispherical ends 5 cm. and 1.5 cm. in diameter, respectively, and a disk 2.5 cm. thick and 8 cm. in diameter, the edges being rounded. Support each of these on a varnished heavy glass rod, as an ignition-tube, set into a suitable base, or suspend each of them by a silk cord. Now charge the first insulated conductor with electricity, and compare the relative magnitude of the charge at different parts by observing how many charges brought by the proof-plane to the electroscope from one place are necessary to produce the same amount of divergence of the leaves of the electroscope as one or more brought from another place. In like manner, test the other conductors. Is there any connection between the electrical density and the degree of curvature of the surface? Try a conductor pointed at one end, made so by inserting a piece of needle. Can you charge it? Explain. Why avoid points on electrical apparatus?

In applying the proof-plane, always place its disk flat against the conductor to avoid changing the shape of the conductor under examination.

329. Exercise. — Determine the effect of neighboring conductors on the distribution of the electrification of an insulated conductor.

Charge with electricity the insulated cylinder of the last experiment, and place near one end of it a similar conductor connected by a wire or chain with the gas, water, or steam pipes of the room. Examine, as in that experiment, the insulated cylinder, comparing the distribution with that found when there was no neighboring conductor.

330. Exercise. — Determine whether the distance of neighboring conductors affects the distribution of electricity on an insulated conductor, and also whether any effects are due to the nature of the medium separating the electrified body from these conductors.

Construct three lead or brass disks 15 cm. in diameter, the edges being neatly rounded, and suspend them by silk cords so as to hang parallel; or, better still, mount them on glass supports, securing them to the glass with wax. Connect each outside one with a wire to an electroscope. Charge the middle one with known electricity, and observe the effect on the electroscopes when the disk is midway between the two outer ones. Move the disk nearer to one of the outer ones, and note the effect. Insert a dry unelectrified pane of glass between these two. Try plates of other substances. Inference.

331. Exercise. — Enclose an electroscope in a cage or box made of wire-cloth. Bring any electrified object in contact with this wire covering, and observe the effect on the electroscope. Now connect the cage with the earth by leading a wire from it to the gas-pipe, and again pass electric charges through it, observing the effect on the electroscope. What method of protecting buildings from lightning is suggested by this experiment?

VIII. CONDENSERS.

332. Apparatus. — Tin-Foil, Paper, Leyden Jars, Electroscopes, etc.

333. Exercise. — Cut three pieces of paper 15 cm. square, and four of tin-foil 10 cm. square. Coat the paper with paraffine or shellac varnish. Arrange these pieces in a pile with the foil and paper alternating. Connect with wire the alternate pieces of foil, making two groups (Fig. 128). Join one of these to the prime conductor of an electrical machine, and the other to the earth. After working the machine for a minute, touch the wire leading to the earth to the one leading to the machine. Explain.

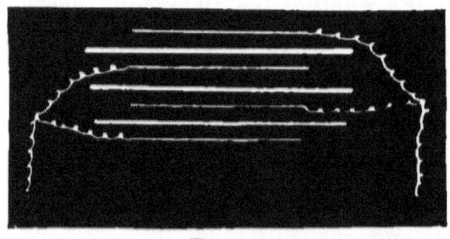

FIG. 128.

The ends of the wires should be rounded. Why? Glass may be substituted for the paper, and gilt-paper for the foil.

334. Exercise. — Place on the table a piece of tin-foil about 25 cm. square, and connect it to the earth by a wire or chain. On this, lay a pane of glass, around which pass two silk ribbons for removing the glass when necessary without touching either the glass or the foil with the fingers. On the glass, lay a small piece of tin-foil, and connect it with the electroscope by means of a wire. Charge with electricity the upper sheet of foil, noticing the effect on the electroscope. Now lift up the glass plate by means of the ribbons, disconnect the lower piece of foil from the earth; then replace the glass plate, all the time watching the indications of the electroscope. Explain.

335. Exercise. — Hold the ball of a **Leyden Jar** (Fig. 129) in contact with the prime conductor of an active electrical machine, at the same time connecting the outer surface of the jar with the negative conductor of the machine by means of a wire or chain. After a few minutes, the jar will have acquired a maximum charge. Using a bent wire, or **Jointed Discharger** (Fig. 130), connect the two surfaces. If the wire is brought in contact with the outer surface of the jar first, no shock will be experienced by the experimenter. After waiting a few moments, you may be able to obtain from the same jar, without recharging, a second discharge of much less intensity than the first, called the **Residual Charge**. The time necessary for the accumulation of the second charge can be shortened by gently tapping the jar with a glass or wooden rod. What change, suggested by this experiment, is probably effected in a body on electrifying it?

FIG. 129.

FIG. 130.

336. Exercise. — Connect the inner surfaces of several Leyden jars by winding a brass chain about the rods passing through the covers. The outer surfaces may be connected by placing the jars side by side on a sheet of tin-foil. Charge the battery in the same manner as a single jar. Compare the intensity of the charge with that of a single jar by the length of the spark seen on discharging.

Leyden jars are easily made out of candy-jars by coating the outside with tin-foil to about two-thirds of its height, using thin flour-paste to cause the foil to adhere. If the opening of the jar is large enough, the inside may be similarly coated; but, if too small to admit the hand, fill the jar two-thirds full with tinsel or crumpled tin-foil. Fit to the jar a varnished wooden stopper, through which passes a heavy copper or brass wire connecting with the inner surface of the jar by a piece of brass chain. The outer end of the brass rod should terminate in a ball. One cast out of lead will answer as well as any. All glass is not equally good as an electric. The electrical qualities of a jar should be tested before pasting on the foil, by tying on the outside a sheet of foil, filling the inside part full of crumpled foil, in which is inserted the brass rod carrying the ball, and then ascertaining if it will retain a charge for several hours.

337. Exercise. — Support a Leyden jar on a pane of glass, and try to charge it by bringing its ball in contact with the prime conductor of an active electrical machine. Compare the intensity of the charge received, with that received in the same time if its outer surface is connected with the other pole of the machine. Account for the difference.

338. Exercise. — Fasten about a Leyden jar a leather or cloth belt into which small carpet-tacks have been driven, with the points extending outward from the glass (Fig. 131), the heads being in contact with the foil. Use this jar as in the last experiment, comparing the charge received when insulated, with that received by the same jar similarly insulated, but without the belt. Explain.

FIG. 131.

339. Exercise. — Charge a Leyden jar so that the inner surface is positive, and a second one so that the inner surface is negative. What will be the electrical condition of the outer surface in each case? Now put the outer surfaces in contact, and see if they discharge. Suspend a pith-ball midway between the balls, and account for its behavior.

340. Exercise. — Charge a Leyden jar having movable metallic coatings (Fig. 132). Separate the jar into its parts, and test the two coatings for electrification. Put the parts together, and then ascertain if there is any charge. What appears to be the office of the coatings?

Fig. 132.

In taking the jar apart, the inner coating should not come in contact with the hand.

IX. ELECTRICAL MACHINES.

341. Apparatus. — Frictional Machine, Holtz Machine, Electrophorus, Proof-Plane, Galvanometer, Electroscope, etc.

342. Exercise. — Make the following tests on a frictional plate-machine : —

1st, Determine the kind of electrification of the prime conductor when the conductor that is connected with the rubbers is joined to the earth by a chain. In like manner, test the combs.

2d, Determine the kind of electrification of the rubbers when the prime conductor is connected with the earth.

3d, Examine the prime conductor for manner of electrical distribution.

4th, Test the upper and also the lower half of the plate for electrification.

5th, Separate the rubbers from the earth, and ascertain if it affects the working of the machine as shown by the striking distance of the spark.

6th, Connect the rubbers with the prime conductor by means of a wire, inserting a short-coil galvanometer in the circuit, and determine, by the indications of the galvanometer, the direction of the current. See Art. 347.

343. Exercise. — Charge the resinous bed of an **Electrophorus** (Fig. 133) by rubbing it with a piece of fur or flannel, and determine the kind of electrification. Place the metal disk in contact with the bed, touch the finger to its upper surface; then remove the finger, lift the disk by its insulated handle, and determine its electrical condition. Try to charge the disk without touching it with the finger. Account for the difference. Explain the action of the machine.

FIG. 133.

To make an electrophorus, fill a pan about 25 cm. in diameter and 2 cm. deep with a mixture of one part, by weight, of beeswax, melted with five parts of shellac. For a cover, cut out of sheet zinc a disk 20 cm. in diameter; turn over the edge neatly, and solder it, forming a rim free from points or sharp edges. Cement to its centre with sealing-wax a piece of glass rod for a handle.

344. Exercise. — Make the following tests on a Holtz machine (Fig. 134): —

1st, Determine the kind of electrification of each armature.

2d, Determine the kind of electrification of different parts of the stationary plate.

MAGNETISM AND ELECTRICITY. 195

3d, Test the combs.

4th, Test the surfaces of the condensers.

5th, Connect the poles of the machine with a wire, and insert a short-coil galvanometer to determine the direction of the current. See Art. 347.

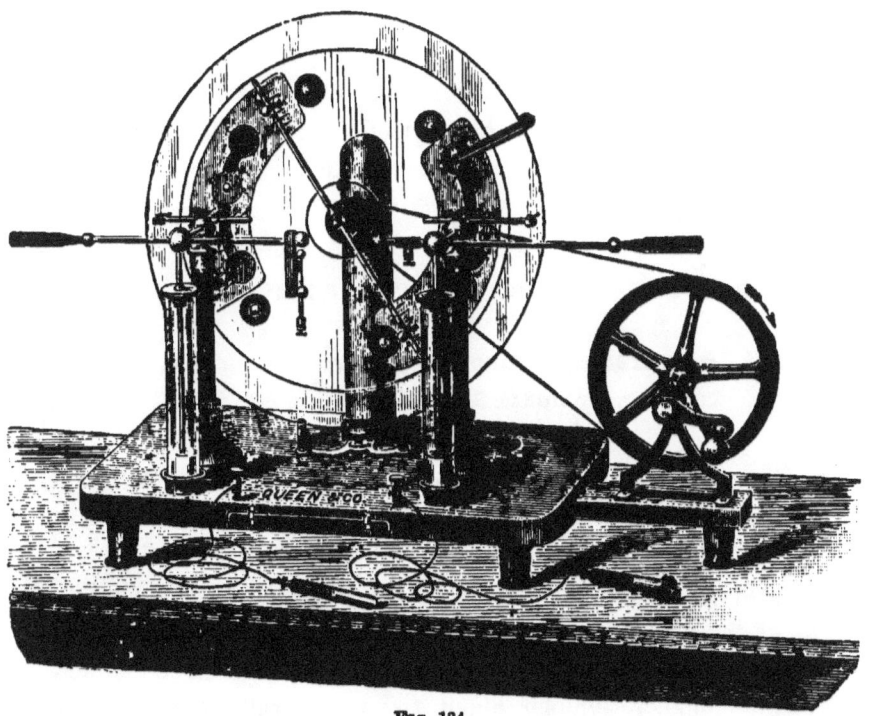

FIG. 134.

6th, Connect the outer surfaces of the condensers, inserting a galvanometer to determine the direction of the discharge; i.e., to determine which one is at the higher potential.

7th, Compare the working of the machine when the condensers are attached, with that when removed.

8th, Determine the effect on the spark produced by exchanging the balls serving as poles, for others of larger diameter.

9th, Examine in a dark room an active machine, comparing the appearance of positively charged points with those which are negative.

10th, Set in rapid rotation, by means of a whirling-machine, a Newton's color-disk in front of the active machine in a dark room. Watch the disk during each discharge to ascertain if it is seen to move while illuminated by the spark. What inference can you make regarding the duration of the spark?

11th, Compare the energy expended in driving the machine when excited, with that when not excited. Whence the electrical energy?

X. VOLTAIC ELECTRICITY. — THE BATTERY.

345. Apparatus. — Zinc, Copper, Iron, Lead, etc., in Sheet Form ; Tumbler, Compass, Insulated Wire, etc.

346. Exercise. — Cut from sheet zinc a strip about 10 cm. long and 3 cm. wide, and from sheet copper a second one of the same size. Fasten to one end of each of these a piece of No. 20 copper wire about 20 cm. long, either by soldering, or by passing the end of the wire through a small hole punched in the strip, and twisted on itself so as to make a good contact. Fill a common tumbler about two-thirds full of water, and add to it slowly about one-twentieth as much sulphuric acid (commercial). Now hold the zinc strip vertically in this dilute acid for a few minutes, observe closely the surface of the zinc, and record any changes. Then place the copper strip in the tumbler, holding it near the zinc, parallel to it, but not touching it, and ascertain whether its presence in any way affects the previously noted phenomenon. Now touch the outer ends of the two strips, or the two attached wires, and

MAGNETISM AND ELECTRICITY. 197

note the effect, if any. Go over these steps several times to make sure that what was seen was no accidental occurrence.

Wash the zinc strip thoroughly in water, rubbing it with a cloth to remove the black particles of carbon from its surface. It may be found necessary to scour the zinc with sand. Apply with a cloth kept for the purpose a little mercury, performing the operation on a common earthen plate. But little pressure should be used in applying the mercury, to avoid breaking the zinc. Remove all jewellery from the hands before handling mercury. Replace the zinc in the dilute acid, and repeat all the tests which were made with the unamalgamated zinc, comparing each observation with the corresponding one previously made.

Ascertain the effect on the phenomena attending the strips, when in the acid, of connecting the two wires with a short strip of cotton, instead of touching their ends. Try successively a piece of silk, wood, paper, gilt-paper, glass, tin-foil, silver, wax, etc.

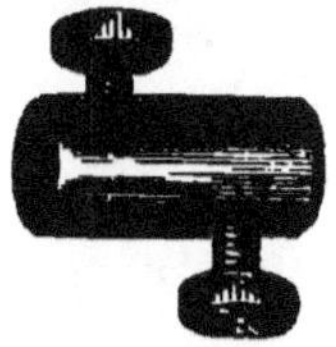

If mercury should get on the copper strip, remove it by heating the strip in a flame. Instead of joining the ends of the wires together by twisting them, it is preferable to employ Connectors (Fig. 135). The connection could be made by dipping the ends into a small cup of mercury.

Fig. 135.

347. Exercise. — Prepare strips of zinc and of copper as in the last experiment. Clean the zinc thoroughly in dilute sulphuric acid, and amalgamate it by rubbing it with mercury. Tack these strips on opposite sides of a piece of wood one centimetre wide and about 12 cm. long, being careful that the tacks from opposite sides do not touch. Place these strips in a tumbler two-thirds full of dilute sulphuric acid. Connect the free ends of the wires, and then stretch a portion of the wire over a compass-needle, or a magnetic needle mounted as in

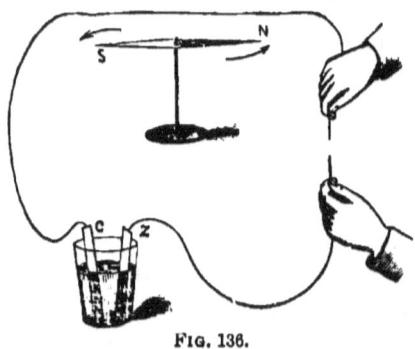

FIG. 136.

Fig. 136, holding it parallel to it. Record the effect produced on the needle. Compare the effect with that produced by placing the wire below the needle. Reverse the wire without changing the position of the plates, and compare the effect with that previously obtained.

Assuming that there is an electric current flowing through the wire from the copper to the zinc, express as a law the behavior of the needle on passing the current parallel to it. See if it is possible to hold the right hand so that the index finger will point out the direction of the current, while, at the same time, the thumb will point out the direction of the needle deflection.

FIG. 137.

Fig. 137 shows a form of apparatus known as **Oersted's Parallelogram**, that is well adapted to the work just described.

FIG. 138.

Ascertain the effect on the needle of passing the current entirely around it (Fig. 138) by bending the conducting wire into a rectangular form, and placing the needle within it. Try two turns of wire about the needle. Try several. Inference.

A simple method of placing a magnetic needle within a

looped conductor, as required above, is to set a common pocket-compass in a hole cut in a square piece of board 15 mm. thick, and a little longer on the edge than the diameter of the compass. Around this board, wind insulated copper wire (No. 20), as many layers as needed. The ends of the wire can be fastened by driving short brass wire staples into the block. This will be found a very serviceable galvanometer for many of the experiments that follow. Binding-posts screwed into the block will make it easier to place it in a battery-circuit. Double connectors may be used instead.

To facilitate reversing the current, as required above, a device called a **Commutator** may be used. One may be made as follows: —

Bore two holes one centimetre deep, and one centimetre apart, in a block of wood. In these pour mercury, connecting each to a binding-post screwed into the block by a short heavy copper wire. The galvanometer is to be connected to the binding-posts, the battery to the mercury-cups. To reverse the current, it will be necessary simply to cross the battery-wires.

Test the rule for determining the direction of a battery-current by means of a galvanometer, by having some one connect it to a battery-cell so placed that you do not see the order of connection made; then, by observing the direction of deflection of the needle, determine which wire is joined to the zinc plate.

348. Exercise. — Make a voltaic element out of strips of zinc and sheet-iron after the plan given in the last experiment. Place the plates in dilute sulphuric acid, connect the poles to the galvanometer, and determine which is the positive one. Try iron and copper; lead and copper; zinc and lead; iron and lead.

XI. EFFECTS OF ELECTRICAL CURRENTS.

349. Apparatus. — Battery, File, Galvanometer, Apparatus for Electrolysis, Fine Wire, etc.

350. Exercise. — Twist the wire leading from one of the poles of a battery about one end of a file, and the other wire about a steel nail. Now draw the nail across the rough file surface. Note the attending phenomenon. Apply the finger to the end of the nail after drawing it several times over the surface of the file, and ascertain if there has been any change in its temperature. Inference.

351. Exercise. — Fasten to each pole of some convenient form of battery a short piece of No. 18 copper wire. Using connectors, join their ends with a piece of No. 30 platinum wire 3 cm. long. Close the circuit, and observe the effect on the platinum wire. Try fine iron wire. Try copper wire. Ascertain if the size or the length of the wire makes any difference. Are wires of different metals affected alike? Place the galvanometer of Art. 347 in circuit with the battery, and take the reading. Now insert the short piece of fine wire in the same circuit, and again read the galvanometer. What is indicated by the decrease in the deflection of the needle? What have you as an equivalent for the loss of electric current?

The battery used must give a strong current. The chromic acid battery is recommended. Any short-coil galvanometer can be used. Be careful to place it so that the wire of the coil is parallel to the needle before closing the circuit.

352. Exercise. — Make a paper tube 5 cm. long and 1 cm. in diameter. Fit corks to the ends. Through one of them put two pieces of No. 18 copper wire, joining their inner ends with a short piece of No. 30 iron wire. Fill the tube with gunpowder, close the ends with corks, and connect the terminals with long wires to the poles of a battery of two or three chromic acid cells located at some distance from the torpedo. Now close the circuit. Explain.

Try a torpedo having the ends of the copper wires not joined by a fine wire, but bent together, leaving a break of about 2 mm. Connect the terminals to the surfaces of a heavily charged Leyden jar. Explain. The **Electrical Mortar** (Fig. 139), or the **Gas-Pistol**, may be used in place of the torpedo to illustrate the heating effects of the electric spark.

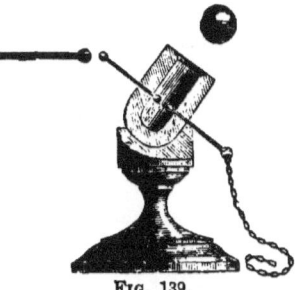

FIG. 139.

353. Exercise. — Bend a glass tube of about 1.5 cm. diameter and 15 cm. long into a V-form. Close the ends with corks and thrust through them platinum wires (Fig. 140), terminating within the tube in strips of platinum foil 2 mm. wide and 3 cm. long, and reaching within 5 cm. of each other. The foil should be soldered to the wires, and varnished at the junction to prevent chemical action on the solder. The tube may be supported in a burette-holder, or by setting it into a slot cut into the side of an empty chalk-box. Fill this V-tube two-thirds full of a solution of sodium sulphate, colored with litmus or the liquor obtained by boiling purple cabbage in water. Connect the

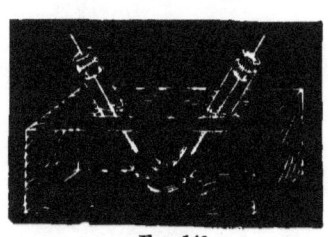

FIG. 140.

terminals to the poles of a battery of two or three cells joined in series (Art. 369). Remembering that the action of an acid on a vegetable purple is to turn it red, and that of an alkali is to give it a greenish tinge, what effect do you find the electric current has on the solution of sodium sulphate? Try copper sulphate. Try tin chloride. Use quite dilute solutions, and omit the coloring matter in the last two cases.

354. Exercise. — Cut the bottom from a wide-mouthed bottle having a diameter of about 8 cm. Insert a good cork in the neck, and through it thrust two platinum wires terminating within the bottle in strips of platinum foil set parallel to each other, and one centimetre apart. Fill the bottle two-thirds full of slightly acidulated water, and support it on the ring of the iron stand. Over these electrodes, invert long test-tubes also filled with the acidulated water. Now place the apparatus (Fig. 141) in circuit with a battery of two or three cells connected in series (Art. 369). Ascertain the relative amount of gases liberated during any interval of time by measuring the length of the column. Remove each test-tube by holding the finger firmly over its mouth; invert, and apply a lighted match. The one that burns with a slight explosion is hydrogen; the other is oxygen.

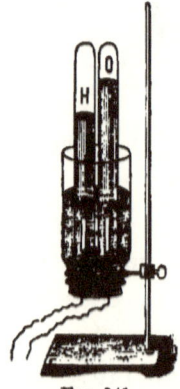

Fig. 141.

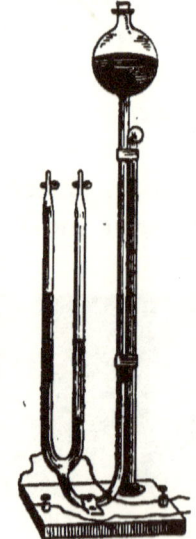

Fig. 142.

Reverse the current through the apparatus, and ascertain if there is any difference in the liberation of the gases.

Ascertain the effect of increasing the number of cells in the battery. Compare two different batteries by observing the length of the hydrogen column produced by each during the same interval of time. A convenient form of apparatus for effecting the decomposition of liquids by means of the electric current is shown in Fig. 142. The gas can be drawn off by means of a rubber tube, and tested. This tube must be filled with water to expel the air. Why? The measurement is readily effected by having the tubes graduated.

355. Exercise. — Dampen a sheet of printing-paper with a solution of ferrocyanide of potassium and ammonium nitrate. Lay it on a brass plate connected with the negative pole of the battery. Connect the positive pole of the battery with a piece of iron or steel wire. Use this wire as a pencil, and write on the prepared paper. Compare the effect with that produced when trying to write with the wire not in the battery circuit. Explain.

356. Exercise. — Wind neatly on a rod of annealed iron about 10 cm. long and 1 cm. or less in diameter, two or three layers of insulated copper wire No. 18 or 20 (Fig. 143). Put the wire in circuit with a battery, and examine the iron core for magnetic properties. Break the

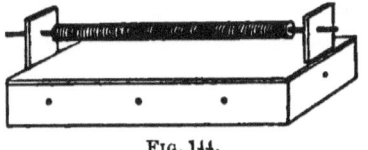

FIG. 144.

FIG. 143.

circuit, and again test the rod. Jar the rod a little, and again test it.

Support the apparatus on a sounding-board (Fig. 144), and listen intently near the coil as the battery circuit is opened and closed. What is the probable origin of the phenomenon? What is suggested by this experiment regarding the nature of magnetism?

XII. ELECTRICAL MEASUREMENTS.

a. Resistance of Conductors.

357. Apparatus. — Galvanometers, Daniell's Battery, Set of Resistance Coils, Wheatstone's Bridge, Insulated Wires, etc.

358. Exercise. — Measure the resistance in ohms, of an electrical conductor.

There will be needed for this work a **Constant Battery**, a **Galvanometer**, and a **set of Resistance Coils**. For most purposes, the Daniell's battery will be found sufficiently constant. Of galvanometers there are a great many forms, each possessing its special points of excellence. In selecting one for any line of work, regard must be had for the conditions under which it is to be used. If the current is feeble, owing to the great resistance of the circuit, a long-coil galvanometer should be chosen in order that a suitable deflection of the needle may be obtained. A galvanometer having two or more coils is to be preferred for general use, as then the resistance can be adapted to the case. A deflection of the needle of between 25 degrees and 55 degrees is preferable.

A serviceable tangent galvanometer is easily made as follows: Cut out of well-seasoned wood a ring 30 cm. in diameter, and having a cross-section of 2.5 cm. square. Mount this on a circular wooden base provided with levelling-screws (Fig. 145). No iron must enter into the construction. In the

outer edge of the ring, cut a rectangular groove 2 cm. wide and 2 cm. deep. In this groove wind two turns of heavy copper wire, the ends to be connected to binding-posts on the base. This will give a coil of practically no resistance. Now wind in
the same groove, using insulated copper wire No. 26, three other coils of 10, 30, and 90 turns respectively, the ends of each to be connected to its own pair of posts. By making connections, so as to secure the sum or difference of these coils, any of the following series can be secured: 10, 20, 30, etc., 130 turns. Fasten across the ring a wooden or brass bar supporting a wooden or paper box 10 cm. in diameter

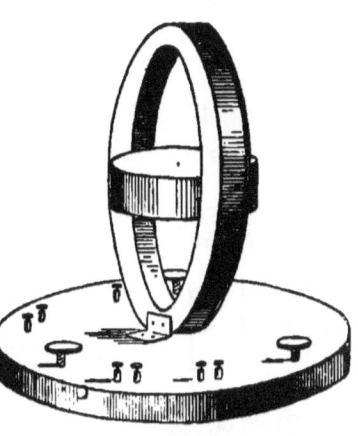

FIG. 145.

and 5 cm. deep, having a glass cover with a hole 1 mm. in diameter drilled through its centre. Place on the bottom of this box a circular piece of mirror-glass to which is cemented a

FIG. 146.

paper ring graduated to degrees. For a needle, magnetize a piece of watch-spring 15 mm. long, inserting it edgewise into a disk of elder-pith (Fig. 146). Cement to the pith-disk, at right angles to the needle, a pointer of glass thread made by drawing out a piece of glass tubing when thoroughly softened in a flame. Suspend the needle

by a silk fibre from the centre of the glass cover. The needle should hang at the centre of the wooden ring, and within a few millimetres of the scale. When the instrument is levelled, the two ends of the pointer should give the same readings, and reversing the current should not alter the amount of the deflection; if it does, the pointer is either not set at right angles to the needle, or the needle is not properly centred. To eliminate small errors of this kind, average the readings of both ends of the pointer; reverse the current, and average the readings; and finally average these two averages for the correct reading.

In Fig. 147 is shown a more sensitive form of galvanometer, where a form of needle known as **Astatic** is employed. Such an instrument is desirable for use in connection with Wheatstone's bridge or the thermopile.

Fig. 147.

For most of the work in this book, a galvanometer constructed as follows will be sufficiently reliable: Procure a circular wooden box 10 cm. in diameter and 10 cm. deep, having a wooden cover with a large circular hole cut in it, exposing the bottom clearly to view. On the bottom, place a wooden block 2 cm. thick, on which is wound a flat coil of several layers of No. 24 insulated copper wire. Bring the ends through the side of the box, and fasten them to small binding-posts. The scale and the needle are prepared as in the case of the tangent galvanometer described above. To the under side of the ring-cover, cement a clean plate of glass. Attach the fibre supporting the needle to the centre of the cover. This must be done with great care, for, if not centred, the two ends of the pointer will give different readings as the cover is turned around. This instrument may be used as a tangent galvanometer, owing to its short needle, without sensible error, for angles not exceeding 45 degrees.

MAGNETISM AND ELECTRICITY.

If the coil is wound in two parts, as shown in Fig. 148, the distance between the parts being a trifle less than the length of the needle, it has been found that the deflection is proportional to the current up to 50 degrees. When no current is passing through the coils, the needle should hang symmetrically between them.

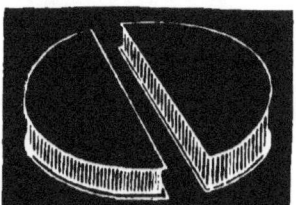

Fig. 148.

Resistance-boxes usually consist of a collection of coils of insulated German-silver wire of different lengths and degrees of fineness, and consequently of different resistances, each coil having some known relation to the standard unit of resistance, the **Ohm.** By means of movable arms or plugs, as the case may be, any required resistance can be thrown into the circuit (Fig. 149): Fig. 150 shows a cheap form of resistance-box suitable for beginners. To close the circuit, each switch must rest on a brass nail-head of the corresponding graduated arc. These contacts must be kept bright to insure good work. In using coils, do not keep the circuit closed

Fig. 149.

Fig. 150.

for any great length of time, as they become heated from the current.

First Method. — Form a circuit consisting of a constant battery B, a galvanometer G, a set of resistance coils R, and the conductor X to be measured (Fig. 151). By disconnecting X, and inserting the short thick wire W, the coil may be cut out of the circuit. *First*, take the reading of G when sufficient resistance has been introduced by means of R to produce a suitable deflection. *Secondly*, insert W, dropping X out of the circuit, and add sufficient resistance through R. to produce the same reading for G. The added resistance is evidently equal to that of X. The deflection of the galvanometer had better not exceed 45 degrees. The accuracy of this method is dependent very largely on the sensitiveness of the galvanometer, and the constancy of the battery. It is more accurate for small resistances than it is for large ones. Be as expeditious as possible in making the required observations, that the conditions may change as little as possible. It is recommended to place a telegraph key in the circuit to make and break connections.

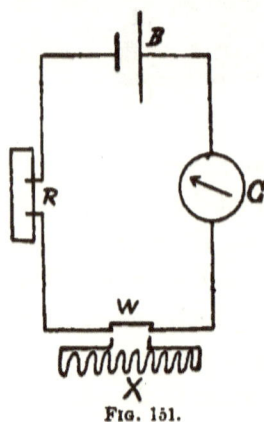

Fig. 151.

Second Method. — Form a circuit consisting of a constant battery B, a tangent galvanometer G, a set of resistance coils R, and the conductor X to be measured, using as little wire as possible in making the connections. Represent the resistance of the battery by r, which, if unknown, can be determined by Art. 367, and that of the galvanometer by g, which, if unknown, can be determined by the first method of measuring resistances of conductors, employing a second galvanometer as a galvanoscope, or as in Art. 365. Adjust G and R so as to secure a deflection between 25 degrees and 55 degrees. With a resistance R and the unknown X in the circuit, take the reading of G which may be represented by a. Now cut out X, and introduce a resistance R' by means of the resistance-box, to give a different deflection of G, as a'. Substitute the values obtained in the formula $\dfrac{R + r + g + X}{R' + r + g} = \dfrac{\tan a'}{\tan a}$, and solve for X.

Proof.—By Ohm's law, $C = \dfrac{E}{R+r+g+X}$, and $C' = \dfrac{E}{R'+r+g}$. By the law of the Galvanometer, $\dfrac{C}{C'} = \dfrac{\tan a}{\tan a'}$. Therefore $\dfrac{R + r + g + X}{R' + r + g} = \dfrac{\tan a'}{\tan a}$.

As the resistance of a battery varies somewhat with the external resistance, it is advisable to employ one having as low a resistance as possible, and, at the same time, make the external resistance large in comparison, so that the effect on the constancy of the battery of changing the external resistance may be small. The best results are obtained by securing a deflection as near as possible to 45 degrees, and having a and a' nearly equal. When the external resistance is large in comparison with the battery resistance, the latter may be neglected with but slight error.

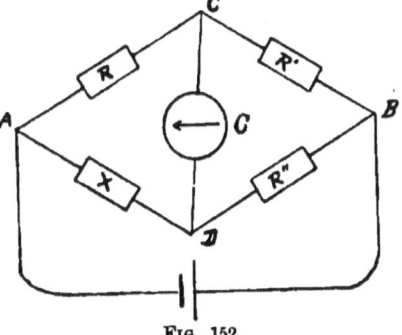

FIG. 152.

Third Method.— For this method is needed a Wheatstone's bridge and an astatic galvanometer. The principle applied is, that, if a circuit be set up as shown in Fig. 152, and R, R', and R'' so adjusted, that, on closing the circuit, no deflection of G is produced, then $R'X = RR''$, whence $X = \dfrac{RR''}{R'}$. Fig. 153 illustrates a simple form of the **Bridge**, in which two of the resistance-boxes are replaced by a fine German-silver wire. This is constructed as follows: On a varnished board 1.1 metre long and 15 cm. wide, fasten with brass screws strips of copper or brass 1.5 cm. wide, of the form shown in the figure, the pieces ME and HF being 10 cm. long. The spaces between A and the cross-pieces may be one

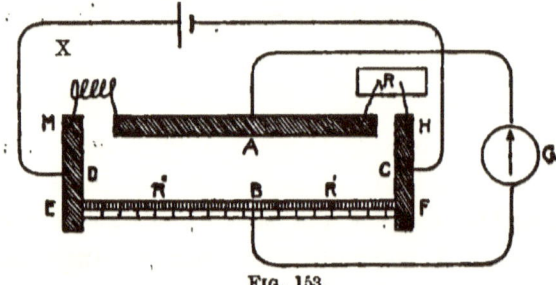

Fig. 153.

centimetre. The lengths of the strips should be such that the inside distance between E and F is exactly one metre. Solder a stretched German-silver wire (No. 26) across from E to F. Under this wire fasten a paper scale divided to half-centimetres. Solder binding-posts to these strips so that connections may be made as shown in the figure, the contact at B being made by pressing the wire from the galvanometer against the wire of the bridge. In making connections, use coarse wire, and as little as possible, that the resistance of these wires may be neglected.

To use the bridge, make the connections as already indicated; through R, introduce 1, 10, or 100 ohms; then slide B along the wire EF till a point is reached at which the reading of G is zero; then X equals the reading of R multiplied by the ratio of EB to BF.

This method is unaffected by any changes in the electrical

current, the degree of accuracy secured depending on the sensitiveness of the galvanometer.

359. Exercise.— Measure the resistance of 40 metres of insulated copper wire, and also of 20 metres of the same kind of wire. Compare the resistance with the lengths. Inference.

360. Exercise.— Measure the resistance of 40 metres of insulated copper wire (No. 23), and also of the same length of No. 30 wire. Compare the resistances with the diameters. Inference.

361. Exercise.— Compare the resistances of equal lengths of wires of different substances having the same diameter. Inference.

362. Exercise.— Compare the resistances of two equal wires of the same material laid side by side, and connected at their ends, with that of one of them taken separately.

363. Exercise.— Measure the resistance of a coil of fine wire of known diameter, but whose length is *unknown*, and also measure the resistance of a *known* length of wire of the same kind. Calculate the length of the unknown coil by the formula $\frac{R}{R'} = \frac{l}{l'}$, R and l representing resistance and length respectively.

364. Exercise.— Determine the effect of change of temperature on electrical conductivity.

In Wheatstone's bridge, for X put one metre of fine wire (No. 30) made into a coil by winding it carefully on a pencil, not allowing the successive parts of the coil to touch, so that the current must pass through its entire length. Now set B so as to produce a balance; then hold a Bunsen flame under the coil; notice the effect on the balance, and ascertain whether the conductivity is increased or diminished by moving B till the balance is restored.

Compare iron, copper, and German-silver wires in this way to see if they are equally affected.

The general effect that the lowering of the temperature of a conductor has upon its conductivity may be roughly shown by connecting the poles of a strong battery with a short piece of iron wire of such a length that the current heats it to a dull red color. Now apply a lump of ice to one portion of the wire, and observe the effect on the other part. Inference.

365. Exercise. — Measure the resistance of a galvanometer by means of its own deflection.

First Method. — Connect, in circuit, the galvanometer, a constant battery of known resistance, and a set of resistance coils, employing short thick wire. Proceed as in Art. 358 (second method), except that the X is omitted. By giving, in succession, two values to R, as R and R', two values for a are obtained, as a and a'. Now substitute in the formula $\frac{R + r + g}{R' + r + g} = \frac{\tan a'}{\tan a}$, and solve for g. It will be necessary to determine the battery resistance unless it is known to be small in comparison with R and g.

Second Method. — Place the galvanometer in the place of X in Wheatstone's bridge, joining the posts A and B (Fig. 153), previously connected by the galvanometer, by a short, thick wire. A deflection of G is now obtained. Now place a bar magnet in the magnetic meridian, so that, by its attraction, the deflection of the needle is reduced to zero. Adjust the resistance in the arms of the bridge, so that the reading of G is not changed by connecting or disconnecting A and B. Then $R'g = RR''$ whence $g = \dfrac{RR''}{R'}$.

b. Resistance of Batteries.

366. Apparatus. — Daniell's Battery, Galvanometer, Resistance Coils, Wheatstone's Bridge, etc.

367. Exercise. — Measure the internal resistance of a battery.

First Method. — Place the cell whose resistance is required in circuit with a tangent galvanometer of known resistance g, and a set of resistance coils. Introduce through the coils a resistance of R and observe the deflection a of G. Then introduce a different resistance R', and observe the deflection a' of G. Substitute in the formula $\dfrac{R + g + r}{R' + g + r} = \dfrac{\tan a'}{\tan a}$, and solve for r. Do not use too large deflections of the galvanometer, as beyond 45° the needle moves out of a uniform field.

Second Method. — When two cells of the same kind are available, then connect them in opposition, that is, with a short thick wire join the two negative poles; and then, by means of their positive poles, place them in circuit with a constant battery, a galvanometer, and a set of resistance coils. Since two equal cells, joined in opposition, neutralize each other, then any deflection of the galvanometer must be due to the constant battery. Hence, two such cells can be treated as a dead conductor, and the resistance measured as in Art. 358. Half of this resistance is that of one of the cells.

It will be found, however, on connecting two cells in opposition and placing them in circuit with a galvanometer, that a feeble current is furnished, showing that the cells are not quite equal. Hence this method is not very reliable.

Third Method. — Place the battery in the place of X in Wheatstone's bridge (Fig. 153), connecting C and D by a wire. Now set R as nearly equal to the resistance of the battery as can be estimated, and find a position for B, such that G maintains a constant deflection whether the circuit is closed between C and D or not. Then $r = \dfrac{RR''}{R'}$. It is better to employ a controlling magnet, and bring the galvanometer reading down to zero, as then the instrument is most sensitive.

The accurate measurement of the resistance of a battery is not easy, as it varies with the resistance in the circuit at the time, and is constantly changing with the chemical changes going on in the cell. The most that should be expected is to get an approximation to its resistance, under the conditions governing the battery during the time it was under examination. Employing high resistances would tend to retard polarization; but their use would be open to the objection that errors in such high resistances will introduce errors into the results obtained larger than the battery resistance. Unless you have the most accurately constructed apparatus it is better to employ resistances but little, if any, greater than those being measured; and, in case the galvanometer is over-sensitive and gives too large a deflection, reduce its deflection by a controlling magnet, or employ a shunt, of known value, and compute the resistance of the shunted instrument, as shown in Art. 376.

368. Exercise. — Determine whether the size of the plates and the distance between them affect the resistance.

Construct a simple cell, as in Art. 346. Measure the internal resistance, when the plates reach 8 cm. into the exciting

MAGNETISM AND ELECTRICITY. 215

fluid, and compare with that when they reach only 4 cm. into the liquid. Also measure the resistance when the plates are 3 cm. apart, and compare with that when 1.5 cm. apart. Inference.

Polarization may be prevented during the experiment by keeping up a gentle agitation of the liquid, by means of a small glass rod.

369. Exercise. — Compare the resistance of two cells, of the same size and kind when joined abreast, with that of a single cell. Also measure their resistance when joined in series. Inferences.

To connect cells abreast, join their like poles together, making of them one pole. To connect them in series, join the positive pole of one to the negative pole of the next, thus leaving the negative pole of the first cell and the positive pole of the last one to serve as poles of the battery.

e. Electro-Motive Force of Batteries.

370. Apparatus. — Same as in the last section.

371. Exercise. — Measure the electro-motive force of a battery.

For this experiment there will be needed a battery giving a constant current whose E. M. F. is known. For most purposes, that is, wherever great accuracy is not required, the Daniell's cell may be used. If the liquids of this battery are solutions of zinc sulphate and copper sulphate of the same density, then its E. M. F. is about 1.1 volt.

First Method. — Connect in circuit with the cell serving as a standard a set of resistance coils and a galvanometer. Now introduce through the resistance coils a resistance of

R ohms and read the deflection a of the galvanometer. Then substitute the cell whose E. M. F. is to be determined for the standard cell, and adjust the resistance R' so as to obtain the same deflection of the galvanometer as before. Hence, as the currents are equal in the two cases, $\frac{E}{R+r+g}=\frac{E'}{R'+r'+g}$, and $E'=\frac{R'+r'+g}{R+r+g}.E$ in which E and E' represent the E. M. F. of the standard cell and the unknown one respectively.

By making R and R' large, then r and r' may be neglected without sensible error, and $E'=\frac{R'+g}{R+g}.E$, that is, the E. M. F. of the two cells are proportional to the resistances which produce equal deflections of the galvanometer. Multiplying the value of E by $\frac{R'+g}{R+g}$ gives the E. M. F. of the cell under examination.

By employing large resistances and a sensitive galvanometer better results will be obtained as polarization is retarded.

Second Method. — Connect in circuit the standard cell, the tangent galvanometer, and the resistance coils. Introduce sufficient resistance R to give a deflection a, something between 20° and 50°. Now substitute for the standard cell, the one of unknown E. M. F., and adjust the resistance R' so that the total resistance of the circuit is the same as when the standard cell was used, and record the deflection a'. This will require that r, r', and g be known. By Ohm's law we have $C=\frac{E}{R+r+g}$ and $C'=\frac{E'}{R'+r'+g}$, whence $\frac{C}{C'}=\frac{E}{E'}$, as the denominators were made equal. By the law of the galvanometer $\frac{C}{C'}=\frac{\tan a}{\tan a'}$. Hence $E'=E.\frac{\tan a'}{\tan a}$. By using a sensitive galvanometer of

MAGNETISM AND ELECTRICITY. 217

several thousand ohms resistance the resistance coils may be omitted and the resistance of the battery neglected.

Third Method. — Connect in circuit with a galvanometer and a set of resistance coils the weaker of the two cells, the standard and the unknown. When a resistance R is used record the deflection a of G. Then add resistance R' to reduce the galvanometer reading any convenient number of degrees, as 10°. The galvanometer should be adjusted to give a deflection in the vicinity of 45°. Now exchange batteries and introduce such resistance R'' as to produce the same deflection a. Then add enough more resistance R''', to reduce the deflection the same number of degrees as before. From Ohm's law it follows that $E' = \frac{R'''}{R'} \cdot E$, that is, the E. M. F. of two cells are to each other as the resistances which produce equal diminutions of current strength.

Proof of the above rule: $C = \frac{E}{R+r+g}$, when a was first observed. On reducing this deflection 10°, then $C' = \frac{E}{R+R'+r+g}$. As the deflection was made the same when the stronger battery was used then the current strength was the same and $C = \frac{E'}{R''+r'+g}$. On reducing this deflection 10°, as before, then $C' = \frac{E'}{R''+R'''+r'+g}$, R''' being the extra resistance required to bring about the necessary reduction. Therefore, $\frac{E}{R+r+g} = \frac{E'}{R''+r'+g}$, and $\frac{E}{R+R'+r+g} = \frac{E'}{R''+R'''+r'+g}$. If the members of these equations are inverted and the resulting equations subtracted from each other, member from member, we have $\frac{R'}{E} = \frac{R'''}{E'}$, whence $E' = \frac{R'''}{R'} \cdot E$.

In measuring the E. M. F. of batteries which polarize rapidly, as the Leclanché, or whose E. M. F. falls rapidly on closing the circuit, as the chromic acid battery, the adjustments required by the above methods must be approximated to at first and the battery then allowed to rest for a few moments to recuperate. After that the circuit may be again closed and the former approximations corrected as rapidly as possible. A few trials will lead to very satisfactory results. It is well to remember, however, that the E. M. F. of a battery changes somewhat with the resistance of the circuit.

372. Exercise. — Determine whether the E. M. F. is affected or not by the size of the plates.

373. Exercise. — Measure the E. M. F. of a cell and compare it with that of two cells of the same kind when joined abreast. Also compare it with that of two when joined in series. Inference.

374. Exercise. — Calibrate any galvanometer so that its readings will be those of a tangent galvanometer.

Connect in circuit a tangent galvanometer, G, the galvanometer to be calibrated, G', a set of resistance coils and a constant battery. Now introduce sufficient resistance to reduce the deflection of G' to 1°, and read G, then lessen the resistance till G' = 2°, and read G, and so on till the equivalents of about 50° have been obtained.

If the resistance is reduced to zero before a deflection of 45° is reached, then increase the current by introducing a second cell, repeating the observations for the last galvanometer deflection observed when the single cell was employed in order that the ratio sustained by the new readings to the old may be obtained and all reduced to the one standard.

If the two galvanometers are so different in sensitiveness that a current which causes a large deflection of one produces but little effect on the other, then connect the terminals of the more sensitive one by a wire, called a *shunt*, so that only part of the current passes through the galvanometer. A galvanometer calibrated in this way must be used in connection with the shunt employed in calibrating it, whenever its readings are to be used in comparison with this particular tangent galvanometer. When used alone the shunt may be dropped and the calibrated readings will be those of a tangent galvanometer.

375. Exercise. — Determine the law for divided electrical circuits, that is, if two conductors join two points of a battery circuit, find how the current in each branch is related to the resistance of that branch.

Set up a circuit as shown in Fig. 154, consisting of four galvanometers, of known resistance, all calibrated by comparison with the same standard, two resistance coils and a constant battery. Short thick wires should be used so that the resistance may be practically confined to the galvanometers and the coils.

1st. Set R and R_1 so that the total resistance in the branches CD and EF are equal. Compare G and G_1, G_2 and G_3, G and G_2. Inference.

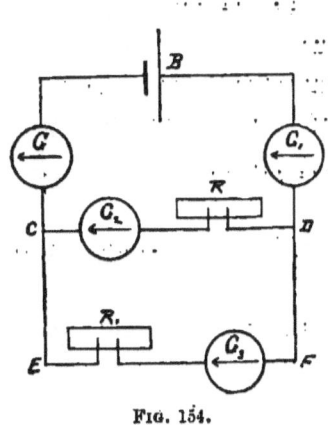

FIG. 154.

2d. Set R and R_1, so that the total resistance in EF is twice that in CD. Compare as before. Inference.

3d. Set R and R_1, so that the total resistance in EF is three times that in CD. Compare as before. Inference.

Frame a law expressing the conclusions reached.

376. Exercise. — Calibrate any galvanometer so that the relative current-strengths are known, and hence the absolute strengths, provided the value in ampères of any one reading is known.

Connect in circuit a constant battery, two sets of resistance coils and a galvanometer (Fig. 155).

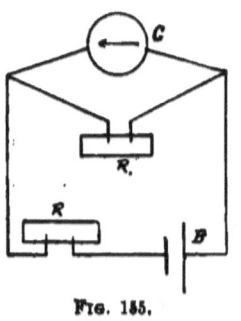

Fig. 155.

If $R_1 = g$, then half of the current goes through G. If $R_1 = \frac{g}{2}$, then a third of the current goes through G. If $R_1 = \frac{g}{3}$, then a fourth of the current goes through G. If $R_1 = \frac{g}{n-1}$, then $\frac{1}{n}$th of the current goes through G.

As the shunt reduces the total resistance of the circuit, to maintain the current constant a suitable compensation must be introduced each time by means of R. To determine the amount of this compensation it is evident that, as the conductivity of the galvanometer and the shunt circuit combined is $\frac{1}{g} + \frac{1}{R_1} = \frac{g + R_1}{g R_1}$, the combined resistance of these circuits is $\frac{g R_1}{g + R_1}$. As $R_1 = \frac{g}{n-1}$, then $\frac{g R_1}{g + R_1} = \frac{g}{n}$. Hence the compensation to be added is $g - \frac{g}{n} = g \cdot \frac{n-1}{n}$.

When $n=2$, $R_1 = g$, the current in G is one-half, and the compensating resistance is $\frac{1}{2}\,g$.

When $n=3$, $R_1 = 2\,g$, current $= \frac{1}{3}$, compensating resistance $= \frac{2}{3}\,g$.

When $n=4$, $R_1 = 3\,g$, current $= \frac{1}{4}$, compensating resistance $= \frac{3}{4}\,g$.

When $n=5$, $R_1 = 4\,g$, current $= \frac{1}{5}$, compensating resistance $= \frac{4}{5}\,g$.

Hence, if we put in the circuit the $\frac{1}{2}$-shunt, and through R the compensation $\frac{1}{2}\,g$, then add resistance through R sufficient to produce a deflection of 1°, on removing the shunt and the compensating resistance the current through G will be doubled.

If this deflection is recorded it will indicate, whenever it is obtained, that the current then flowing is double that which gave the deflection of 1°.

Now, try in succession the $\frac{1}{3}$-shunt, $\frac{1}{4}$-shunt, etc., with the proper compensating resistances, as determined above, and deflections for three times, four times, etc., the current at 1° will be sent through the galvanometer. Tabulate these deflections and the calibrated scale for relative current strengths will be known.

XIII. ELECTRO-MAGNETISM AND ELECTRO-DYNAMICS.

377. Apparatus. — Batteries, Iron Filings, Insulated Wire, Bar Magnet, Electro-Magnet, Ampère App., Magnetoscope, Floating Battery, Sounder, Telegraph Key, Relay, etc.

378. Exercise. — Connect two or more large cells, in parallel circuit, with short thick wire. Close the circuit through a heavy wire, and then dip a portion of it into iron filings. Compare the result with that obtained when the wire is carrying no current.

379. Exercise. — Thrust the wire of the last experiment vertically through a sheet of stiff paper supported in a horizontal position (Fig. 156). After closing the circuit sift a few iron filings on the paper, jarring it slightly with the finger as they fall. Compare the figure with that given when the wire carries no current. Inference.

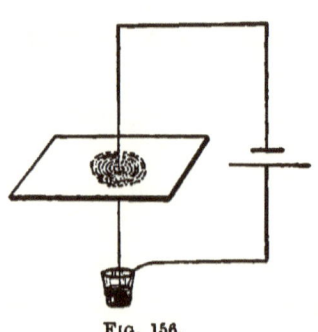
Fig. 156.

380. Exercise. — Coil into flat spirals, 8 cm. in diameter, some No. 20 insulated copper wire, securing the wire in place by means of sealing-wax (Fig. 157). Suspend these from two wire hooks, fixed to a piece of board and connected to binding-posts, the hooks being placed so that the coils hang parallel and about 1 cm. apart. The lower ends of the wires dip into a vessel of mercury to a depth of 1 mm. Now pass a current from a chromic acid, or Bunsen cell, through these coils, by joining the battery-poles to the posts on the board, and record the effect. Turn one of the spirals over, and again record the effect. Trace the direction of the current through the spirals, in each case, and devise a law embodying the results.

Fig. 157.

Straight wires may be substituted for the spirals; then, if the battery-poles are connected to the posts, the current will move in opposite directions in the two wires; and if one battery-pole

MAGNETISM AND ELECTRICITY.

is joined to both posts and the other pole is connected to the mercury-cup the currents will have the same direction. A battery of several cells joined abreast will be necessary when straight wires are employed.

381. Exercise. — Thrust the straight part of the wires of the last experiment through a sheet of paper supported horizontally, and study the magnetic field, as in Art. 379; first, when the current in the wires have the same direction; and secondly, when opposite. Test the effect of increasing the current by increasing the number of cells. How would you join them?

382. Exercise. — Determine the laws of electrical currents.

Construct a rectangular frame, 25 cm. square, out of insulated copper wires, No. 20, by winding about four layers around the edge of a square board. Slip the wire off the board, and tie the parts together in a number of places with fine cord. Bend one end of the wire into a hookform (Fig. 158), the other end to be left straight, extending out from the other side. Bend a narrow strip of brass into a J-form, giving the hooked part a spoon-shape, so that it will hold a small globule of mercury. Support the strip in a bu-

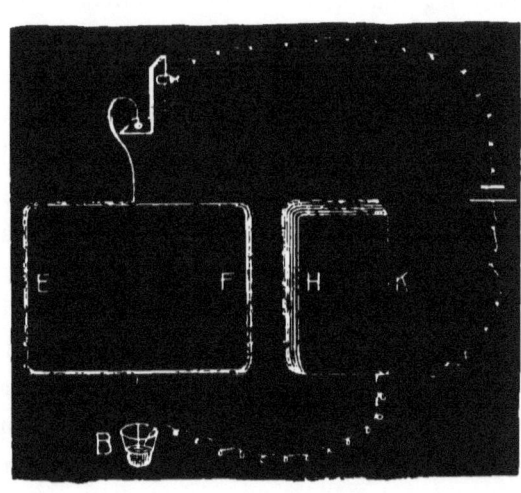

Fig. 158.

rette-holder, and hang the rectangle from it with the straight end dipping into a vessel of mercury. Adjust the shape of the hooked part of the wire so that the upper and lower sides of the rectangle are horizontal, the frame keeping any position given it, and turning on application of the slightest force. The supporting point of the frame should be conical, good contact being secured by a globule of mercury. Connect one pole of the battery to the strip, and the other to the vessel of mercury.

Wind four or five layers of insulated wire around the edge of a board 20 cm. by 10 cm. Connect this in the same circuit with EF, support it near to EF and in the same plane, with the current moving through it parallel to that in F, and in the same direction, and record the effect on EF. Inference.

Remove HK from the circuit, and hold a bar magnet parallel to EF, or place it beneath it and in the same plane, and record the effect. Reverse either the magnet or the direction of the current, and again record the effect. What law seems to govern the phenomenon?

Coil some No. 16 insulated wire into a close spiral (Fig. 159), arranging it for suspension in the same manner as in the case of the wire rectangle. Give the spiral, usually called a **Solenoid**, a diameter of 4 cm. and a length of 15 cm. Suspend the solenoid in place of the wire frame, adjusting it so that it hangs horizontally and turns freely on its support. Close the circuit, and then hold a strong magnet near one of its ends. Determine the effect of reversing the current. Trace the current

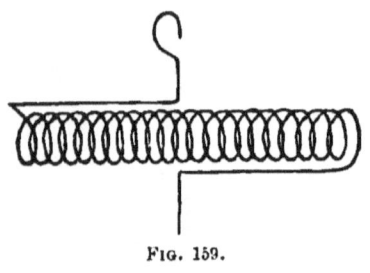

FIG. 159.

through the solenoid. In what direction is it moving in the N-seeking end?

Apply the law derived from the study of the solenoid to an **Electro-Magnet** (Fig. 160) to determine its polarity when in circuit with a battery. Test the conclusion reached by placing it in a battery circuit and bring its poles in succession near a magnetoscope.

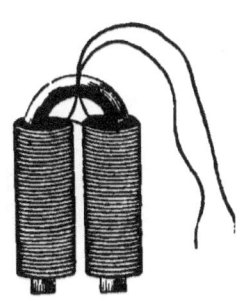

Fig. 160.

Double part of the wire of the circuit back on itself as shown in Fig. 161, and hold it parallel to one of the vertical sides of the suspended wire rectangle. Account for the lack of motion on the part of the rectangle. Use a strong current.

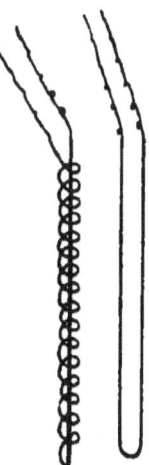

Fig. 161.

Substitute for the simple loop a spiral, with the return wire passing in a straight line through it (Fig. 161). Compare the result with that obtained in the last case. Is the action of the sinuous portion any greater than that of the rectilinear? What must be the effect on the solenoid to have the wire bent parallel to its axis, as shown in Fig. 159?

Mount a magnetic needle so as to be free to move in any direction in a horizontal plane. Hold parallel to the needle, and a little above it, a wire carrying an electric current of known direction, and record the effect. Compare this with the solenoid when an electric current passes parallel to its axis, above it, and in the same direction with respect to the line joining its poles, as it did in the case of the magnet. Try the current below in each case. Ascertain the effect of reversing the current.

383. Exercise.—Make a small chromic acid battery, using a short wide test-tube, or a light bottle, for a battery-jar. Insert

Fig. 162.

it in a cork of sufficient size to float the apparatus in a vessel of water (Fig. 162). Connect the poles of the battery with a small helix of wire. Float the apparatus in a vessel of water, and observe if it assumes any one position in preference to another. Hold a straight wire forming part of a battery-circuit parallel to the axis of the helix. Record the effect. Hold a pole of a permanent magnet near one end of the helix. Record the effect. Frame a law expressing the action of electric currents on each other, as well as the action of currents on magnets, as proved by this floating battery.

If the edges of the cork-float are greased, there will be less tendency on the part of the apparatus to keep against the side of the vessel, especially if the latter is filled to the brim with water.

384. Exercise.—Set up a telegraph line, the longer the better. A return wire may be used, or the wires may be connected to the gas-pipes, if rooms in the same building are joined. If distant buildings are connected, the earth may be used for the return circuit. In that case earth connections may be secured by connecting the wires to the gas or water pipes, if possible, or to large metal plates buried deep enough to reach moist earth. If the line is set up within a room, a coil of German-silver wire may be put in circuit to make the line resemble a long one. In any case the insulation of the line must be secured. Measure the resistance of each instrument and also of the complete line. Connect a sounder, a relay, and a set of resistance coils in the line, and compare their working as the

XIV. CURRENT INDUCTION.

385. Apparatus. — Galvanometer, Battery, Helices of Wire, Electric Motor, Bar Magnet, Microphone, Coils of Insulated Wire, Sheets of Rubber, Copper, Paper, Zinc, and Iron, Induction Coil, Telegraph Key, etc.

386. Exercise. — Wind smoothly a few layers of insulated copper wire, No. 24, on a hollow paper or wooden cylinder with thin walls (Fig. 163). Connect this helix with a sensitive galvanometer. Now introduce suddenly into the helix the N-seeking pole of a strong bar magnet. Why is the needle deflected? What direction must the current have had? Is there any current when the magnet is stationary within the coil? Suddenly remove the magnet and compare the effect with that previously obtained. Try the S-seeking pole. Try magnets of different strengths. Try a helix having less wire. Conclusions.

FIG. 163.

387. Exercise. — Place within the helix used in the last experiment a soft iron core, consisting of a number of No. 18 iron wires, the length of the spool. Now move the pole of a strong bar magnet suddenly up to the end of this core, and observe the effect on the galvanometer. Try moving the pole of the magnet rapidly away from the core. Trace out the currents produced in the helix, and compare their direction with the Ampèrian currents of the magnet.

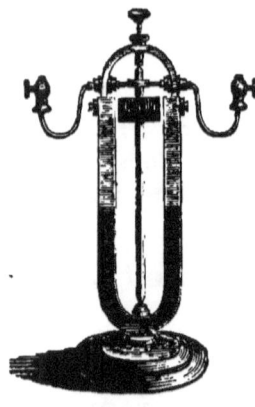

Fig. 164.

388. Exercise. — Connect an Electric Motor (Fig. 164) to a sensitive galvanometer of low resistance. Rotate the armature of the machine by rapidly pulling off a string wound on the shaft. Observe the effect on the galvanometer. Reverse the direction of rotation. Explain the action of the machine in the light of the laws of induction developed in Art. 386.

389. Exercise. — Wind on a paper bobbin, 2 cm. long, about 30 g. of No. 30 insulated copper wire, and place it around the pole of a bar magnet, supported horizontally (Fig. 165). Connect the terminals of the helix to a sensitive galvanometer. Cut out of thin sheet-iron, such as photographers use in "tin-

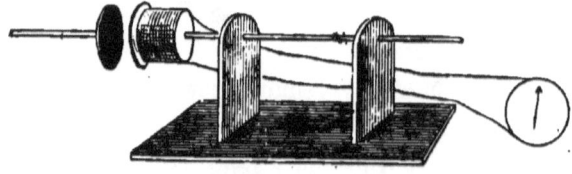

Fig. 165.

typing," a disk 8 cm. in diameter. Cement to it a wooden handle. Now move the disk suddenly toward the pole of the magnet and observe the effect on the galvanometer. Try moving it suddenly away from the pole. Explain.

Replace the galvanometer by a second bobbin and magnet (Fig. 166). Opposite the pole of the second magnet and near

MAGNETISM AND ELECTRICITY. 229

to it suspend, by a thread, an iron disk to the centre of which is cemented a piece of mirror. Place the apparatus where a ray of sunlight may strike the mirror and be reflected on the wall. Now move the first disk in the same manner as

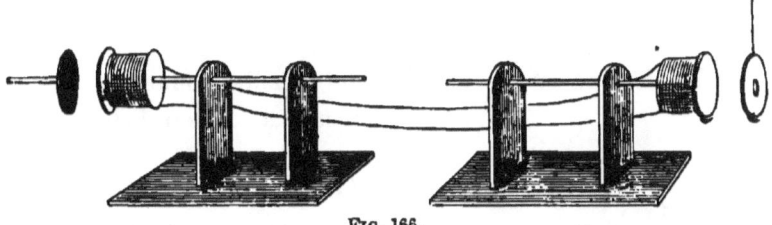

Fig. 166.

when the galvanometer was in circuit, and observe the effect on the second disk by means of the spot of light on the wall.

The apparatus will be more sensitive if the edge of the disk is opposite the pole of the magnet, for then it will turn on a vertical axis.

Bell telephone receivers may be substituted for the magnets and bobbins employed above.

390. Exercise. — Fasten two square pieces of gas-carbon to an upright strip of thin wood, and join them by a carbon pencil with tapering ends, resting loosely in conically shaped cavities in the horizontal bars (Fig. 167). Place this **Microphone*** in circuit with a battery and a Bell telephone by

Fig. 167.

---

* The Microphone does not illustrate Current Induction. Its consideration at this point is found convenient on account of using it in connection with the Telephone.

twisting wires around the horizontal bars. Listen through the telephone to the ticking of a watch resting on the base of the microphone. Connect two rooms electrically, the microphone and telephone being in the circuit, but in different rooms. Listen at the telephone to the talking of an assistant standing near the microphone.

391. Exercise. — Insert within a helix connected with a galvanometer a second helix in circuit with a battery (Fig. 168). Observe the effect of introducing or removing suddenly this helix. Try opening or closing the circuit without moving the helix. Set up the apparatus with a set of resistance coils in the battery circuit. Ascertain the effect of weakening or strengthening the current by varying the resistance. Reverse the direction of the battery current and repeat the tests. Write a law describing the direction of the induced currents as compared with the battery current.

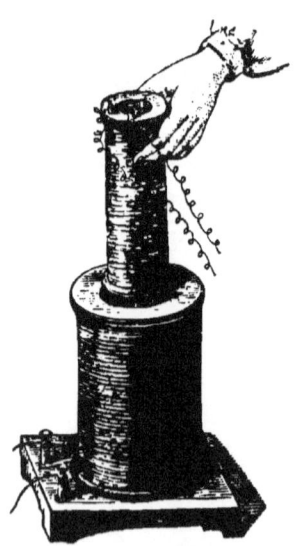

FIG. 168.

392. Exercise. — Construct two large flat coils of insulated copper wire No. 24. Place one of them in circuit with a battery and some form of current breaker, as a telegraph key, and the other in circuit with a sensitive galvanometer. Lay one of these coils on the other on the table, with a sheet of cardboard between them. Read the galvanometer on closing the circuit and also on opening it. Substitute a sheet of india-rubber for the paper. Try a sheet of copper. Try a sheet of zinc. Try a sheet of iron. Inference.

393. Exercise. — Connect one pole of a battery with a file, and draw across its rough surface the wire attached to the other pole. Observe the intensity and the color of the sparks. Repeat the experiment after introducing into the circuit a helix of wire. Explain. Place an iron core in the helix, and repeat. Explain.

394. Exercise. — Connect the secondary coil of a small induction coil with a galvanometer, and the primary coil with a battery (Fig. 169). Read the galvanometer on closing the circuit and also on opening it. Repeat the observations after removing the iron core. Explain.

FIG. 169.

Remove the galvanometer, and test the differences in the induced currents by the physiological effects experienced on holding the wires in your hands.

The helices of Art. 391 may be used for the above experiment.

395. Exercise. — With a battery, a long coil of wire, a sensitive galvanometer, and a telegraph key set up a circuit as shown in Fig. 170.

On closing the circuit the needle is deflected by a part of the current passing through the galvanometer. Now bring the needle back to zero, and place some impediment in the way so it cannot move in that direction, but is free to swing in the opposite. Break the current, and account for the needle's moving in the opposite direction to that in which the battery current would cause it to move.

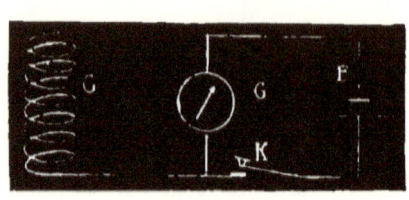

Fig. 170.

396. Exercise. — Connect one pole of the secondary coil of an induction coil, giving a spark at least 2 cm. long, to one surface of a Leyden jar, and the other pole to the other surface. Observe the effect on the spark given by the coil.

397. Exercise. — Around the edge of a wooden barrel-hoop wind several layers of No. 24 insulated wire, fastening the wire in place by tying it in several places with cords. Leave the ends of the wire quite long, and connect them with an astatic galvanometer of low resistance. Now hold the coil in both hands in a plane at right angles to the direction of the dipping-needle, and quickly rotate it through 180° around a horizontal axis lying east and west. Observe the effect on the galvanometer. Try the effect of rotating the coil in the opposite direction. Explain.

Ascertain if there is any induction when the axis of rotation is horizontal and in the magnetic meridian. Try the axis parallel to the dipping-needle.

If such a hoop is mounted on an axis provided with a crank for rapid rotation, the wires being connected to a commutator on the axis, quite marked electrical effects are easily obtained.

XV. LUMINOUS EFFECTS.

398. Apparatus.—Tin-Foil, Panes of Glass, Plates of Mica, Aurora-Tube, Air-Pump, Glass Goblet, Geissler's Tubes, etc.

399. Exercise.—Cut with a small punch a number of circular pieces of tin-foil, and paste them on a pane of glass, a

Fig. 171.

fish-globe, a glass tube, or a Leyden jar without an outer coating, forming any desired pattern (Figs. 171, 172). The edges of these disks should be about 1 mm. apart. Connect the poles of an induction coil or of an electrical machine with the extremities of the design and pass electric sparks through it, conducting the experiment in a dark room.

400. Exercise.—Paste on a plate of mica about 15 cm. by 10 cm., a piece of tin-foil 8 cm. by 5 cm. Support it on a plate of glass with the mica surface

Fig. 172.

234 PRACTICAL PHYSICS.

out, and bring the poles of an electrical machine or induction coil within 1 cm. of the surface, the distance between them not exceeding the length of the tin-foil. Set the machine in operation, darken the room, and observe the path of the spark across the mica plate.

401. Exercise. — Substitute for the mica plate of the last experiment a plate of glass, covered with iron-filings, and having a narrow strip of tin-foil pasted across the two ends beneath the poles of the machine. To make the filings adhere, coat the plate with mucilage, sift on evenly the filings, and set away to dry.

402. Exercise. — Exhaust the air from the **Aurora-Tube**

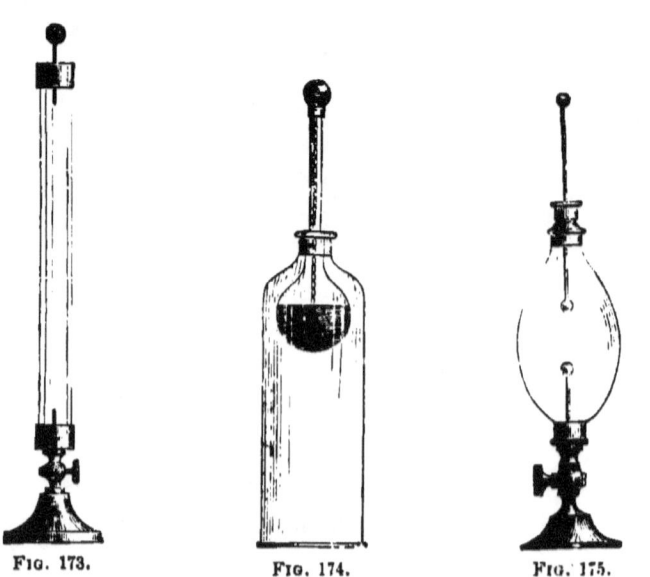

FIG. 173. FIG. 174. FIG. 175.

(Figs. 173, 175), and connect the extremities to the poles of an active electrical machine or induction coil. Study its appearance in a darkened room. Note the effect of reversing the current.

Substitute for the aurora-tube the apparatus shown in Fig. 174, where the mouth of a bell jar is closed with a Florence flask, on which is pasted tin-foil, covering two-thirds of the bulb. A brass chain hangs within the flask, touching the inner surface. One pole of the electrical generator is connected to the brass chain, and the other to the plate of the air-pump. Observe the effect in a dark room, as the air is exhausted from the bell-jar.

403. Exercise.—Coat a glass goblet inside and outside with tin-foil (Fig. 176) a little over half-way up, and place it on the table of the air-pump, beneath a bell-jar provided with a brass sliding-rod passing air-tight through the cork of the jar. Over that part of the rod, within the jar, secure a glass tube, adjusting the apparatus so that the rod touches the bottom of the inner surface of the goblet. Connect the rod and the air-pump table to the poles of an active induction coil or Holtz machine, and then exhaust the air. Conduct the experiment in a dark room.

FIG. 176.

The best effects are given by a goblet made of uranium glass.

404. Exercise.—Exhaust the air from the aurora-tube (Fig. 175), introduce a gas, as oxygen, hydrogen, coal-gas, etc.; then exhaust once more, refill and again exhaust, repeating the operation for several times, in order that the tubes may contain nearly pure gas. Now proceed as in Art. 402.

Dealers in physical apparatus furnish vacuum tubes, or Geissler's tubes, containing rarefied vapors of known substances for the study of the effects of the electrical discharge on the residual gas.

XVI. THERMO-ELECTRICITY.

405. Apparatus. — Strips of Zinc, German-Silver, Copper, Bismuth, Antimony, etc., and a Short-Coil Astatic Galvanometer.

406. Exercise. — Prepare two strips, one of zinc and the other of German-silver, each 5 cm. long and 15 mm. wide. Solder them together at one end, and to the free ends solder wires, so as to place them in circuit with a sensitive galvanometer. A short-coil galvanometer, having an astatic needle, is to be preferred. Apply the flame of a burning match to the junction of the two strips and observe the effect on the galvanometer. Ascertain the effect of a piece of ice on the junction.

Try strips of copper and zinc; copper and bismuth; antimony and bismuth. Compare the deflections given.

XVII. SECONDARY OR STORAGE BATTERIES.

407. Apparatus. — Sheet-Lead, Red-Lead, Galvanometer, etc.

408. Exercise. — Cut out of sheet-lead two rectangular strips, each 10 cm. by 12 cm., and tack them on opposite sides of a strip of dry wood 2 cm. thick. Solder a wire to each strip, then dip the plates into dilute sulphuric acid, and connect them to the poles of a battery of sufficient E. M. F. to decompose water. After a few minutes disconnect the plates from the battery and join them to a galvanometer or small electric motor. Inference. Ascertain whether the plates will exhibit any electrical phenomena or not, if some little time is allowed to elapse between disconnecting them from the battery and testing them with the galvanometer.

409. Exercise. — Cut out of thin sheet-lead a number of plates, 10 cm. by 15 cm., having on one side an ear 2 cm. by 8 cm. (Fig. 177). Cut out of cotton flannel twice as many rectangular pieces, 14 cm. by 20 cm. Make a thick paste out of red-lead and sulphuric acid, and apply it with a brush to both sides of each lead plate, and also to the nap-side of the flannel, to a depth of at least 1 mm. Now build up a battery as follows: Lay down a piece of flannel, then one of lead, then flannel, then blotting-paper of the size of the flannel, then flannel, then a lead plate with the ear on the reverse side of the last plate, then flannel, and so on. Always place the nap-side of the flannel next to the lead, and be careful to have as many plates with the ear on one side as there are plates with the ear on the other, consecutive plates having their ears on opposite sides. When a sufficient number have been put in place, bind them together with a cord, strips of thin board being placed on the outsides. Wire together the ears forming the two poles, and place the plates in a glass vessel containing dilute sulphuric acid. To charge the battery, connect the poles to those of a strong battery for an hour or so. On disconnecting the poles and joining them to an electric motor even after the lapse of several hours a rapid motion should be obtained.

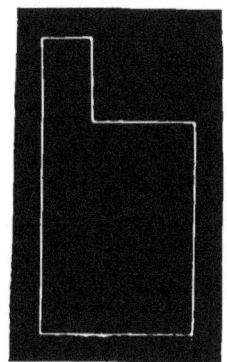

Fig. 177.

CHAPTER VI.

SOUND.

I. WAVE MOTION.

410. Apparatus. — Balls, Whirling-Machine, Spiral of Spring Wire, Cotton Rope, Rubber Tubing, Rectangular Trough, Tin Tube, etc.

411. Exercise. — Classify the following motions in two ways, after studying them closely to ascertain their points of resemblance and of difference.

1st. A ball suspended by a thread, and set in motion as a pendulum.

2d. A heavy ball suspended by a spiral spring, made by winding closely on an iron rod, 1 cm. in diameter, some spring brass wire No. 18, and set in motion by pulling down the ball and then releasing it.

3d. A heavy ball, suspended by a wire fastened to it, so as not to turn on it as an axis, and set in motion by twisting the wire.

4th. A paper disk clamped to the axis of the whirling-machine, and put in motion by turning the drive-wheel.

5th. A weight with a cord attached, and put in motion by drawing it along on the table.

6th. A wire stretched tight, with its ends secured by winding them around the heads of two stout nails driven part way into a board, and put in motion by pulling the middle of the wire a little out of line and then releasing it.

SOUND.

412. Exercise. — Place a soft cotton rope, 5 m. or 6 m. long, in a straight line on the floor, fastening one end to some heavy object. Holding the other end of the rope in the hand, set up vibrations in it by a quick movement of the hand up and down. Ascertain the path traversed by a point of the rope. Compare the motion of consecutive points. Draw figures in your note-book showing the form of the rope at three successive points of time. Ascertain the effect of changing the time in which the hand makes its up-and-down movement in starting the vibration. Account for the curved form's advancing along the rope.

413. Exercise. — Fasten one end of the rope used in the last experiment, or, what is better, a piece of rubber tubing, having an outside diameter of 1.5 cm., to a hook in the ceiling of the room, and the other end to a hook in the floor directly beneath it. Adjust the rope so that it is under a slight tension. Strike the rope with a short rod at one-fourth from the bottom, and note the action of the pulse formed. Just as this pulse starts to return from the ceiling start a second one, and observe the effect attending its meeting the first one. Try striking the rope one-sixth from the floor. Draw figures showing the form of the rope at the moment the pulse was started; on its arrival at the ceiling; on its starting to return; and at the time the two pulses meet.

414. Exercise. — Fasten up two similar ropes or tubes as in the last experiment, subjecting them to equal tensions. Start pulses in them simultaneously, and of the same length, and compare the rates at which they advance along the ropes. Now increase the tension of one rope, and repeat the comparison. What force is developed by stretching the rope? Fasten up a third rope, much heavier than the others; a rubber tube filled with sand will answer. Compare the movement of pulses in this heavy one with similar pulses in the others. Account for the difference. Increase the tension of the heavy rope, and repeat. Derive from the observations made what determines the speed of propagation of the pulse.

415. Exercise. — Fill about two-thirds full of water a rectangular trough, 1 m. long, 30 cm. deep, and 10 cm. wide, having one side of glass. By suddenly depressing and elevating a large bottle in one end of the trough start a wave, and record the movement of the water as it appears through the glass side. Start other waves, and observe their behavior toward one another. Ascertain if one wave can be started so as to increase the amplitude of the one that preceded it. Try to reduce the amplitude. Record the conditions you had to observe in order to accomplish these results.

Float some small paraffine balls on the water, and record their motion as the pulse travels along. In the advance of the wave is there an advance of the water? Load some of the paraffine balls with iron-filings, so that they will float at some distance below the surface. Determine from the motions of the balls the character of the motion of the water particles when a wave traverses the trough.

416. Exercise. — Construct a spiral spring, 3 m. long, by winding closely on a rod, 1 cm. in diameter, a quantity of No. 18 spring brass wire. Fasten one end of this spiral to a hook in the wall; and, holding the other end in the hand, stretching the spiral till it swings clear of the floor, send a pulse through it by pressing a few of the coils of the wire close together, and suddenly releasing them. Tie short pieces of string into several of the coils, and observe their movements as the pulse advances. Describe the motion of the parts of the spiral. Representing the successive coils by a series of short parallel lines, show by a figure the change in relative position of these lines in the neighborhood of a pulse.

417. Exercise. — Place on a table a tin tube, 3 m. long and 10 cm. in diameter, one end of which tapers to a diameter of 2.5 cm. It is preferable to have the tube in parts, to be put together like stove-pipe. Tie over the large end a paper membrane, and in front of the small end place the flame of a lighted candle. Observe the effect of slapping two books together in front of the paper membrane. Account for the effect produced on the flame. Why could it not have been due to a wind produced by the books?

II. SOURCES OF SOUND.

418. Apparatus. — **Bell, Tuning-Fork, Long Metal Rod, Pane of Glass, Glass Rod, Tin Flageolet, Tin Whistle, Glass Tube,** etc.

419. Exercise. — Strike a large bell or glass bell-jar with a small mallet, made by inserting a stout wire in a large cork. Suspend by a thread a small ball of pith or cork, so as just to touch the edge of the bell. What do you find to be the difference between a sounding bell and a silent one?

420. Exercise. — Insert two large screw-eyes in a board at points distant one-half of a metre. Stretch between them a piece of steel wire, about No. 20, securing tension by turning the screws. Now draw a violin-bow across the wire, and record any attending phenomenon. Ascertain the condition of the wire by touching it with the edge of a card. Inference.

421. Exercise. — Tap a tuning-fork against a block of wood, and, while sounding, touch one prong to the surface of a vessel of water. In what condition are the prongs of the fork?

422. Exercise. — Clamp, in a horizontal position, a metal rod, 1 m. long and 1 cm. diameter, by its middle in a vise. Suspend by a thread a small ball, ivory or glass, about 1 cm. in diameter, so as to rest gently against one of the squared ends of the rod. Rub the other end of the rod longitudinally with soft leather, covered with powdered rosin. What does the effect produced on the ball indicate regarding the condition of the rod?

423. Exercise. — Balance on a small cork a pane of window-glass, about 25 cm. square, and sprinkle evenly over it a little fine sand. Rest on the glass plate, in a vertical position, over the centre of the cork, a stout glass tube 60 cm. long. Hold this tube by the upper end in one hand, press it firmly against the plate, and rub the lower part longitudinally with a damp woollen cloth held between the thumb and the forefinger of the other hand. What phenomena attend the friction of the cloth on the tube? Are they connected? Inference.

424. Exercise. — Connect to a faucet giving water under considerable pressure a tin flageolet, by means of a rubber tube slipped over its mouth, but not covering the embouchure. It will be necessary to wrap the tube with wire to keep it from slipping off. Place the flute in a tall vessel of some kind, and turn on the water. By properly adjusting the water-supply a low musical tone can be obtained on the flute's becoming immersed in water. Touch the vessel with the hand while the

SOUND. 243

flageolet is sounding, comparing the sensation with that experienced when the flageolet is silent. Inference.

425. Exercise. — Insert a short wooden or tin whistle in one end of a glass tube about 30 cm. long and about 2.5 cm. in diameter, closing the other end with a cork (Fig. 178). A suitable whistle may be obtained at any toy store. Distribute within the tube a spoonful of fine cork-dust, obtained by filing a cork. Hold the tube in a horizontal position and blow the whistle. Observe the effect on the cork-dust. What must be the condition of the air within the tube when the whistle is sounding?

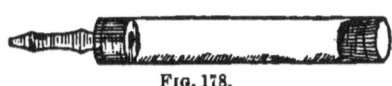
Fig. 178.

Remove the whistle, place the mouth to the end of the tube, and sing a prolonged note of considerable intensity. By trial a tone can be found which will produce the same effect as that given by the whistle.

426. Exercise. — Draw out a glass tube to a jet (see page 356). Connect it by rubber tubing to the gas-supply, support it vertically in a clamp, light the gas escaping from the jet, and reduce the flame to the height of about 2 cm. Clamp

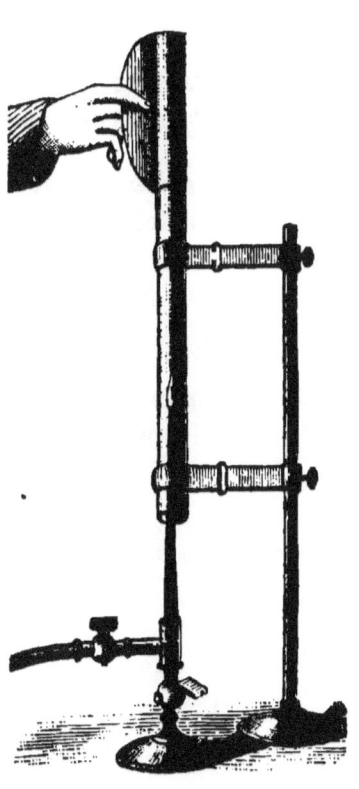
Fig. 179.

over the jet a glass tube with thin walls, about 40 cm. long and 2 cm. in diameter (Fig. 179). Change the position of the jet within the tube till a musical note is emitted by the tube. Compare the flame within the tube when sounding with the flame when the tube is silent. Grasp the lower end of the tube with the hand, and compare the sensations produced by the sounding tube with those produced by the silent one. Oscillate a small mirror back of the flame, and compare the image of the flame when the tube is silent with that given when the tube is sounding. Explain.

III. TRANSMISSION OF SOUND.

427. Apparatus. — Air-Pump, Alarm-Clock, Watch, Rubber Tubing, Tuning-Fork, Bar of Iron, etc.

428. Exercise. — Place under the receiver of an air-pump a small alarm-clock, separated from the plate by a bed of cotton-wool. Compare the distinctness with which the sound of the alarm-bell is heard before and after exhausting the air. Repeat the experiment, omitting the cotton-wool. Inference.

429. Exercise. — Place a loud-ticking watch on one end of a long strip of lath, and hold the other end to your ear. Compare the distinctness with which you hear the sound with that when the ear is away from the lath. Ascertain the effect of wrapping up the watch in a cloth, and then laying it on the lath. Inference.

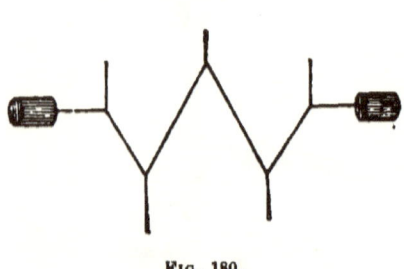

FIG. 180.

430. Exercise. — Connect two rooms by a wire or string

stretched between them (Fig. 180). Support the wire, wherever needed, by passing it through a loop made of cord. The wire must not touch any objects, nor turn any sharp corners. Connect each end to the bottom of a collar-box or tin cylindrical can, by passing it through a small hole, and fastening it into a button. Let an assistant talk near the open end of one box, while you listen at the other. Explain.

431. Exercise. — Fill a piece of rubber tubing, say 2 m. long and 2 cm. in diameter, with water, supporting the two ends so that the tube hangs in a curve. Set a tuning-fork in vibration, and compare the distinctness with which you hear the sound, as an assistant holds the stem in the water of the tube at one end, while your ear is as close as possible to the other end, with that when the tube contains only air. Inference.

432. Exercise. — Suspend a small bar of iron or steel by two strings, each 1.5 m. long. Take one string in each hand, and, stooping over, press the ends into the ears. Now let an assistant strike the swinging bar with a block of wood. Compare the distinctness of the sound heard through strings of cotton, linen, metal, etc. Inference.

IV. VELOCITY OF SOUND.

433. Apparatus. — **Tuning-Fork, Cylindrical Jar, Glass Tube, Rods of Glass and of Wood**, etc.

434. Exercise. — Hold a vibrating tuning-fork, whose rate is known,[*] over the mouth of a cylindrical jar, about 35 cm. high and 5 cm. in diameter, and pour in water slowly, till a point is reached where the sound is greatly strengthened.

---

[*] A fork whose rate is from 250 to 300 will be found to give the best results. If a fork of higher pitch is used, then select a cylindrical jar of less diameter than 5 cm.

Ascertain, by repeated trials, the air-column giving the loudest resonance, and then determine its length. To the length of the air-column add two-fifths of the diameter of the cylinder, then four times this sum will be the wave-length for this fork. Multiply this by the vibration-number of the fork, and the product will be the velocity of sound in air, at the temperature of that within the cylinder. Why?

Ascertain the effect that lowering the temperature has upon the velocity of sound, by packing the jar in melting ice, and determining the air-column required for the best resonance, and hence the velocity of sound in this cooled air. Record the temperature of the air.

A form for keeping the record is given below.

	Length of resonating jar at	
	Temperature of room ....C.,	Tem. of jar in ice ...C.
First observation . . .	................cm.	cm.
Second " . . .	................cm.	cm.
Third " . . .	................cm.	cm.
etc.	etc.	etc.
Mean	................cm.	................cm.
Wave length	................cm.	................cm.
Velocity of sound. . .	................cm.	................cm.
Change in velocity for 1° C.................cm.		
Vibration-number of tuning-fork used		

435. Exercise. — Determine the velocity of sound in carbonic acid gas.

As in the last experiment, obtain the air-column giving the loudest reënforcement of the tone of a tuning-fork whose vibration number is known. Now fill the jar with carbonic acid, and then add water till the reënforcement is the greatest. Proceed as in the last experiment.

Compare the ratio of the velocity of sound in carbonic acid

gas and in air with that of the densities of these substances, and determine how the density of the medium affects the velocity of sound. Tabulate results as in the last experiment. The moment when the cylinder is full of gas can be determined by inserting a burning match; the flame will be extinguished on reaching the gas.

For method of preparing carbonic acid gas, consult Shepard's Chemistry, Exp. 102.

436. Exercise. — Determine the velocity of sound in hydrogen or coal gas.

Close gas-tight one end of a glass tube 1 m. long and 4 cm. wide. Fit a cork into the other end, to slide with slight friction, using a long stout wire for a piston-rod. Mount the tube in a vertical position, mouth downward, with the piston pushed up to the closed end. Now fill the tube with hydrogen (see Art. 102), by displacement; and then by moving the piston determine by trial the position where the loudest resonance is obtained for a vibrating tuning-fork of known rate. Then proceed as in Art. 434. What does this experiment show is the effect of density on the velocity of sound?

437. Exercise. — Determine the ratio of the velocity of sound in a rod of glass, wood, or metal to that in air.

Close one end of a glass tube 1 m. long and 4 cm. wide with a cork, A (Fig. 181). To a rod 1.5 m. long of the material in which the velocity of sound is to be determined, as EB, glue a thin cork piston, B, of such a size as to slide freely within the tube

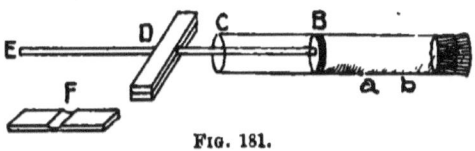

Fig. 181.

AC. By means of an iron clamp fasten the rod at its middle to the table, between two pieces of wood, having the shape shown at F; the thickness being such that the rod is in the axis of the glass tube AC as it rests on the table; the hole through these pieces being just large enough to hold the tube without crushing. As AC is free either to move toward D or away from it, the space AB can be lengthened or shortened at pleasure. Distribute a little cork or silica powder evenly between A and B. Now excite longitudinal vibrations in BE by gently stroking it toward E with a damp cloth, if the rod is glass, or with soft leather covered with powdered rosin, if it is wood or metal, being careful not to spring the rod out of line for fear of breaking it. These vibrations will be communicated by the piston B to the air between A and B, and disturb the powder. By moving AC, a position for B can be readily found at which the agitation of the powder is greatest, being thrown into clearly marked heaps of uniform length subdivided into parallel ridges. Find the average length of these dust heaps, ab, by dividing the distance between A and B by the number of heaps. Representing this length by s, $2s$ will be the length of the air-wave. If the length of the rod is d, since it is clamped at the centre, then $2d$ is the wave-length of the vibration produced in it. Hence $d \div s$ represents the number of times that sound travels faster in the rod than in air at the temperature of the room. Represent by V the velocity of sound at the temperature of the room, t. Then $V = 333 (1 + .0037\ t)^{\frac{1}{2}}$, and the velocity of sound in the rod $= \dfrac{d}{s} V$.

SOUND. 249

V. PROPAGATION OF SOUND.

438. Apparatus. — Cardboard, Whirling-Machine, etc.

439. Exercise. — Cut out of stiff cardboard a disk having a diameter of 30 cm. Concentric with the circumference of the disk draw a circle with a radius of 2.5 mm. Divide the circumference of this small circle into twelve equal parts (Fig. 182), numbering them 1, 2, 3, etc. With the point 1 as a centre, and with a radius of 7 cm., draw a circle. With 2 as a centre, and with a radius of 7.3 cm., draw a second one.

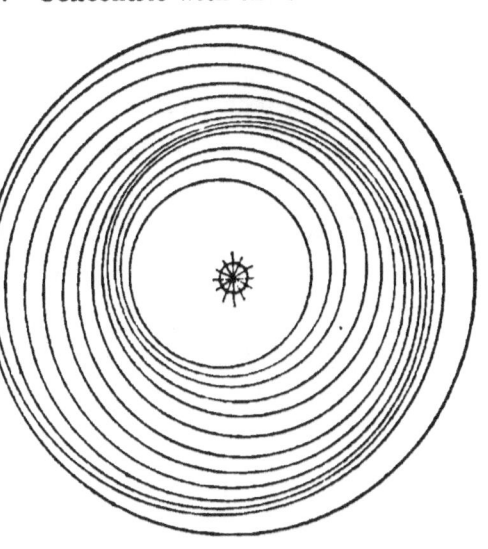

Fig. 182.

With 3 as a centre, and with a radius of 7.6 cm., draw a third one. Continue in this way, increasing the radius each time by 3 mm., till each point has been used twice as a centre. The lines should be drawn with jet-black ink, and should be ½ mm. in width. Clamp this disk on the spindle of the whirling-machine, and hold in front of it, and as close as possible, a strip of cardboard in which is cut a slit 3 mm. wide and 10 cm. long. In this opening will now be seen parts of these circles appearing as black dots unevenly distributed. Now rotate the disk slowly, and waves of condensation, followed by waves of rarefaction, will appear to move along the slit.

Watch closely one dot, and from its motion account for the effects produced. If these dots represent the air-particles along the radius of a series of sound waves set up in air, then in their motion you have a representation of the movements of the air-particles in the propagation of sound waves.

VI. REFLECTION OF SOUND.

440. Apparatus. — Cardboard, Whirling-Machine, Watch, Concave Reflector, Tin Tubes, etc.

441. Exercise. — Cut out of cardboard a disk having a diameter of 30 cm. Out of opposite sides of the disk (Fig. 183) cut two sectors of about ten degrees. Clamp the disk to the spindle of a whirling-machine, and as the disk is made to revolve, blow a whistle or toy trumpet in front of it, and close to it. Compare the sound with that when the disk is at rest. When is the sound loudest? When lowest? Explain.

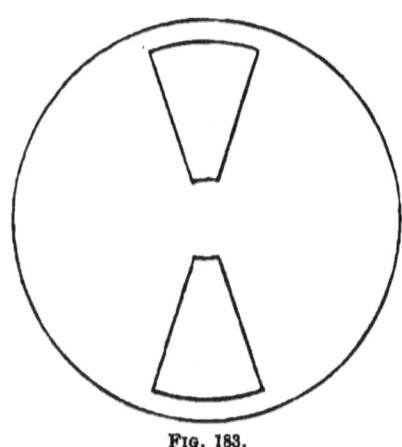

FIG. 183.

442. Exercise. — Suspend a loud-ticking watch at the focus of a large concave reflector, such as is used behind wall lamps. About two metres from this place a second reflector, parallel to the first, and with its concave surface toward the watch. At its focus support a small funnel, with its mouth toward the reflector, and a rubber tube leading from its stem to your ear. Compare the distinctness with which you hear the ticking of the watch with that when the reflectors are removed. Explain.

The focus of the reflector can be found by letting sunlight fall upon it, and marking the point in front of it at which it will set fire to a match.

By suspending the watch a few centimetres farther away from the reflector than the focus, a point can be found by trial, still farther removed from the reflector, at which the watch can be heard with much greater distinctness than at any intervening or more distant points.

443. Exercise. — Determine the law of reflection of sound. Lay on a table two tin tubes, each 1.5 m. long and 10 cm. wide,

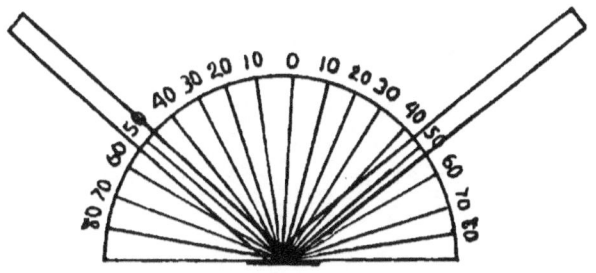

Fig. 184.

forming a V (Fig. 184). Suspend a loud-ticking watch in the outer end of one of the tubes, and listen for the sound at the outer end of the second tube. On a sheet of cardboard draw a semicircular protractor of at least 30 cm. radius, dividing it into 10 sectors. Let these tubes rest on radial lines, on opposite sides of the zero, so that the angle each tube makes with the zero radius can be read off on the scale. Across the other ends of the tubes, and perpendicular to the zero radius, hold a piece of cardboard. Move the tube to which the ear is applied to a position where the ticking of the watch is most distinct. Compare the angles between the zero radius and the tubes. Make several trials for different positions of the tubes. Tabulate the results. Inference.

VII. REFRACTION OF SOUND.

444. Apparatus. — **Rubber Balloon, Watch,** etc.

445. Exercise. — Fill a small rubber toy balloon with carbonic acid by tying it over the delivery-tube of a carbonic acid generator. Suspend a loud-ticking watch from a suitable sup-

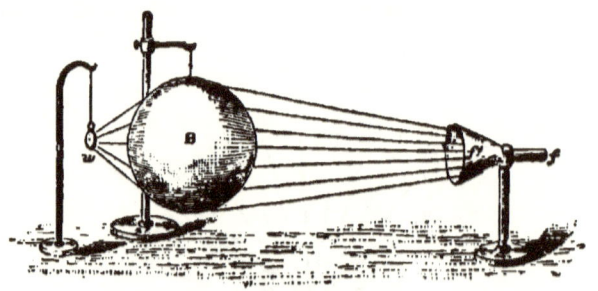

Fig. 185.

port, and move as far away as possible and still hear the watch when one ear is turned toward it. Now hold the balloon close to the ear, and ascertain if the loudness of the sound is affected. Explain.

Fig. 185 illustrates another method of conducting this experiment. The ear is placed at *f*.

446. Exercise. — Fill a balloon with hydrogen gas, and proceed as in the last experiment. Explain. Try a balloon filled with air. Is the effect the same?

VIII. LOUDNESS OF SOUND.

447. Apparatus. — Tuning-Fork, Sonometer, Bell, Air-Pump, Tin Tube, Steel Rod, Glass Tubes, Fruit-Jar, etc.

448. Exercise. — Compare the vibratory movement of a feebly sounding tuning-fork with that of a loud sounding one, by means of the effects seen on touching one prong to the surface of water. What is here shown to be the cause of the difference in loudness?

449. Exercise. — Compare the vibratory movement of a feebly sounding string on a violin or sonometer with that of the same string when loudly sounding, by suspending in contact with it a small paraffine ball. Account for the difference in loudness.

450. Exercise. — Ascertain if the loudness of sound is affected by the density of the medium enveloping the vibrating body.

1st. Observe the change in the sound emitted by a small bell under a bell-jar of an air-pump, as the rarefaction is increased. Inference.

The bell should stand on a bed of cotton-wool. Why? The bell-jar should be provided with a sliding-rod for ringing the bell (Fig. 186).

2d. Fill a large bell-jar with hydrogen, by displacement over water. With one hand lift the jar vertically upward, and with the other ring a small bell within it. Compare the sound with that when the jar is full of air. Inference.

FIG. 186.

Invert the jar, fill it with carbonic acid gas, and repeat the test. Inference.

451. Exercise. — Lay on a table the tin tube of Art. 417. Hang a watch in the opening at the large end, place your ear at the other end, and compare the loudness of the ticking with that when the tube is removed. Explain the action of the tube.

452. Exercise. — Set a tuning-fork in vibration, and compare the sound, when held in the hand, with that when its stem is held firmly in contact with the bottom of an empty chalk-box. Ascertain what is the condition of the box when the sounding fork is in contact with it, by noticing the effect on a little fine sand sprinkled over it. To what is the increased loudness due?

Find out whether the fork vibrates as long when its stem rests against the box as when held in the hand. In making this comparison, blows of equal intensity must be given to the fork. A simple method of doing this would be to let a wooden ball swing against the prong, falling in each case from the same height.

If you have two forks of the same pitch, remove them from the sounding-boxes, placing the boxes on the table in front of you. To give both forks the same impulse, cut out of a board about 6 or 8 mm. in thickness a piece of the shape shown in Fig. 187, the width of each of the prongs, A, A, being somewhat greater than the normal distance between the prongs of the tuning-fork. Slip the latter on obliquely, and straighten up at right angles to the plane of the board, with the surface of the latter even with the surface of the ends of the prongs. Arrange the other fork in the same manner, with respect to the other prong of the board. Now let an assistant hold the handle of the board and strike it a sharp blow with

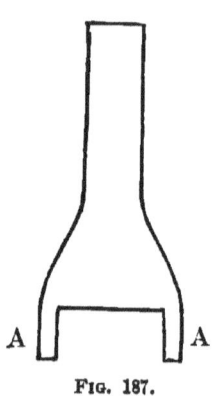

FIG. 187.

a hammer, at a point near the forks, while you hold the latter, one in each hand, prongs downward. Both forks will then be put in vibration with nearly the same amplitude. Place one on its sounding-box immediately, holding the other in the hand. As soon as the sound of the one on the box ceases to be heard, place the one held in the hand on its box. Repeat the experiment, reversing the positions of the forks, to determine if the difference is due to the forks. Why does the energy of one fork hold out longer than the other?

453. Exercise. — Compare the loudness of the sound emitted by a vibrating steel rod, clamped by one end in a vise, with that produced by a steel bar of the same length and thickness, but having considerable width. Account for the difference. Compare the duration of the sounds. Why is there a difference?

Why is a tuning-fork usually made with two branches?

454. Exercise. — Procure four glass tubes, each about 30 cm. long, having different diameters, as 1.5, 2.5, 3.5, and 4.5 cm., respectively. Clamp each one successively in a vertical position, with one end dipping into a vessel of water. Hold over the open end a sounding tuning-fork, raising or lowering the tube till a position is reached where the sound bursts out with maximum loudness. Measure the air-column in each case, compare it with that given by computation for the fork employed, and deduce the correction to be made for the diameter. Tabulate the results.

455. Exercise. — Paste a piece of paper over the mouth of an empty glass fruit-jar. Cut away the paper cover, little by little, the part removed having the form of a segment, till a loud resonance is obtained on holding a vibrating tuning-fork close to the opening. Ascertain the cause of the resonance by sprinkling a little fine sand over the paper, and watching it as the sounding-fork approaches the opening.

456. Exercise. — Compute the length of the air-column 2 cm. in diameter that will reënforce a fork, say C', at the temperature of the room. Then cut eight glass tubes of this length and diameter (see page 354), grinding the ends smooth on a piece of sandstone wet with spirits of turpentine. Cut a number of paper disks, of the same diameter as the outside of these tubes, and also a number of rubber connectors. Now it is evident that a tube whose length is any multiple of the first can be prepared by joining these together with the rubber connectors, and that by cementing a paper disk over the end of a tube an open one can be changed into one closed at one end.

Prepare open tubes whose lengths are respectively 2, 3, 4, etc., times the length of the first, and determine by trial which reënforces the sound of the fork used in preparing the tubes.

Prepare and test closed tubes in the same way.

If we represent by *one* the length of the shortest open tube which reënforces the sound, what will be the lengths of the different open tubes found to reënforce it?

Prepare a similar series for closed tubes.

IX. INTERFERENCE OF SOUND.

457. Exercise. — **Tuning-Fork, Sonometer, T-tubes, Rubber Tubing, Glass Tubing,** etc.

458. Exercise. — Hold a vibrating tuning-fork near the ear, and as you rotate it slowly about the axis of the stem, mark the positions where the sound is feeblest, and also those where loudest. Now hold the same fork over a cylindrical jar, serving as a resonator; find by trial the position in which the sound is feeblest; then cut off the waves sent out by one prong, by sliding over it a paper cylinder, without touching it, and observe the effect. Explain the varying loudness of the sound as the fork rotates about the axis of the stem.

459. Exercise. — Mount on resonant boxes two large tuning-forks of the same pitch (Fig. 188). Stick a piece of sealing-wax, the size of a hickory nut, on the end of one branch of one of the forks. Set the forks in vibration, and compare the sound with that emitted by them before they were loaded. Ascertain the effect of using a lighter or heavier load. Explain.

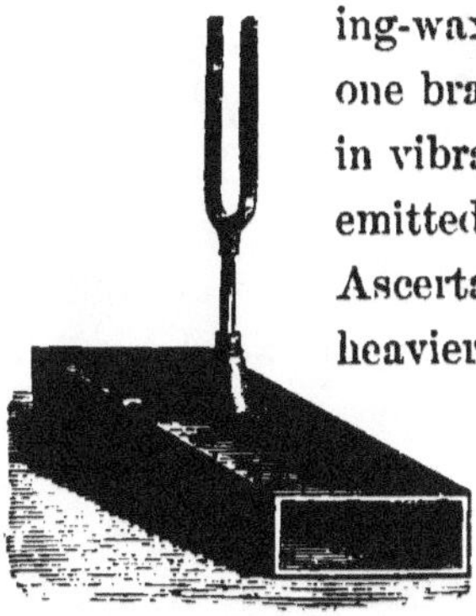

A resonant box is a rectangular box of pine, open at one end. Its length is one-fourth of a wave-length of the fork, less about two-fifths of the diagonal of the end of the box.

Fig. 188.

460. Exercise. — Tune two strings on the sonometer to the same pitch, and then compare the sound emitted by them when plucked simultaneously with that when the tension of one string has been slightly changed.

461. Exercise. — Connect two large T-tubes with rubber tubing, as shown in Fig. 189. Join a funnel to the stem of one of them, and a piece of rubber tubing to the other. Hold a vibrating tuning-fork at A. Now it is evident that the waves divide at B, part traversing each branch, and unite again at K, reaching the ear through H. If the tube CDE is half a wave-length longer than F, the waves will meet in opposite phases at K, and no sound at H will be heard. Similar effects will follow if the difference is $\frac{3}{2}$, $\frac{5}{2}$, $\frac{7}{2}$, etc., of a wave-length. If CDE is a wave-length or some multiple of a wave-length longer, no diminution of intensity will be noticed. Any differences in length other than these will result in partial interference.

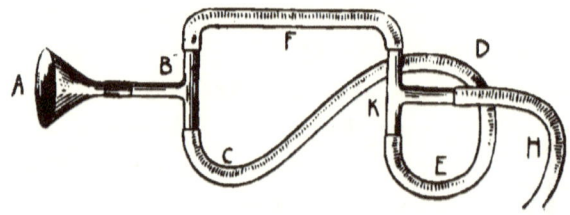

Fig. 189.

Friction of the waves on the walls of the tubing will produce slight deviations from the above results, and the sound may not be completely extinguished owing to the conductivity of the material.

462. Exercise. — Set up two singing flames (see Art. 426), employing tubes of the same length and diameter. Make a paper cylinder to slide with a little friction over the end of one of the tubes so that its length may be increased if desired. Adjust the lengths of the tubes by moving the paper cylinder till the tones are alike; then note the effect of changing the position of the paper cylinder.

X. SYMPATHETIC VIBRATIONS.

463. Apparatus. — **Pendulums, Bar Magnets, Tuning-Forks, Glass Tubing, Sonometer,** etc.

464. Exercise. — Suspend a heavy weight by a wire or string 1 m. long, and find the time of vibration. Now try to

SOUND.

put this pendulum in vibration by puffs of air directed against the centre of the weight, and timed in unison with the pendulum. Try puffs of air irregularly timed. Try puffs of air whose period is some multiple of that of the pendulum.

Arrange a second pendulum to swing in the same time as the first. Connect the bobs of the two pendulums by a string, which is drawn straight when the pendulums hang vertically. Now set one of the pendulums in vibration, in the plane of the two, and note the effect on the second one. Shorten one of the pendulums by lowering the suspension-point, keeping the bobs in the same horizontal line, and ascertain if the effect is the same as before. Substitute a stiff, very light bar for the string between the bobs, and observe the effect on the time of vibration.

465. Exercise. — Suspend two pendulums, each one metre long, from a horizontal bar; a third, one-fourth of a metre long; a fourth, one-half of a metre long; a fifth, 90 cm. long. Set one of the first two in vibration, and observe the effect on the others. Explain.

466. Exercise. — Construct two pendulums, using bar magnets for bobs. Suspend each magnet by two threads, each 50 cm. long, so that it hangs horizontally. Place the magnets parallel, unlike poles adjacent and several centimetres apart. Now set one pendulum in vibration in a plane perpendicular to the other magnet, and study the motions for some time. Explain. Shorten one of the pendulums, by lowering its supporting-point, without changing the position of the magnet, and ascertain if the effects are the same as before.

467. Exercise. — Place near each other on the table the tuning-forks of Art. 459. Set one of them in vibration, and after waiting a few minutes stop it and listen near the other one to ascertain whether it is motionless or not. Would the effect be the same if one of the forks were surrounded with a paper cylinder? What would be the effect of changing the pitch of one fork by sticking a piece of sealing-wax to one of its prongs? Why?

468. Exercise. — Tune in unison two strings on a sonometer (Fig. 192), or a violin. Set one string in vibration; then after a few moments dampen it with the hand, and ascertain if the second one is silent. Shorten one string by moving the bridge under it and repeat the experiment. Explain.

469. Exercise. — Prepare a singing-flame as in Art. 426. Now shift the position of the jet within the tube by lowering it just enough to stop its singing, and then sound near it the same note as was emitted by the tube at first. Can you account for the effect on the flame? Will singing a higher or lower tone work as well? Try it.

470. Exercise. — Prepare a jet-tube by holding one end of a piece of glass tubing in a gas-flame till the opening has been reduced to about 1 mm. wide. Connect this by a rubber tube to the gas-supply, and regulate the flow till a flame some 30 cm. or more in length is obtained, and just on the point of flaring. Now blow a shrill whistle, rattle a bunch of keys, or hiss, and the long flame will be shortened nearly half, and rustle loudly. Turn your back toward the flame and repeat the sounds. Repeat the experiment, standing in a distant part of the room. Explain.

If the gas-pressure is not sufficient to give a flame of the character described, fill a large rubber bag with gas, connect

SOUND. 261

the jet-tube to it, and apply pressure to the bag by means of a heavy weight.

XI. PITCH OF SOUND.

471. Apparatus. — Siren, Cylindrical Jar, Tuning-Forks, etc.

472. Exercise. — Measure the number of vibrations made in a second by a vibrating body.

First Method. — Cut out of cardboard a disk 25 cm. in diameter. Draw on it four concentric circumferences, having the diameters 23, 21.5, 20, and 18.5 cm. respectively (Fig. 190). Divide these circumferences into 48, 36, 30, and 24 equal parts respectively. At these points of division, with a sharp, hollow punch, having a diameter of about 1 cm., cut round holes. Mount this disk on the spindle of a whirling-machine, or on the armature-shaft of a small electric motor, and you have a form of **Siren**.

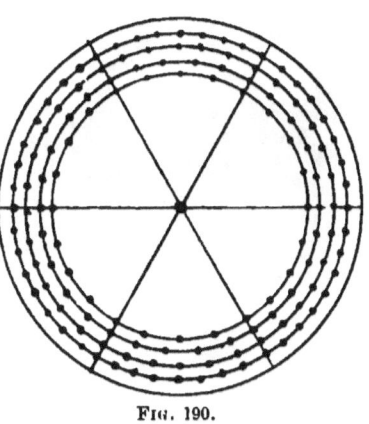

FIG. 190.

If, with a rubber tube, you direct a steady stream of air against one of these rows of holes as the disk rotates rapidly, a musical note is produced by the air-puffs passing through them. Find what the pitch of the tone depends upon.

Now give the disk such a speed that the tone emitted by blowing steadily against one of the circular rows of holes is in unison with that produced by the body in question. Count the number of revolutions made by the drive-wheel in, say, 20 seconds, the speed being kept constant. Find the relative size of the drive-wheel and the spindle-pulley, and from the data compute the number of revolutions made by the disk in a second. This multiplied by the number of holes in the circle will give the vibration-number sought.

The speed of the armature of the electric motor may be determined by clamping in contact with it a speed-indicator, such as machinists use in ascertaining the speed of shafting.

In Fig. 191 is shown a form of siren with a counter. The perforated disk is given a horizontal position, and made to rotate by the stream of air forced through the obliquely cut openings by means of a bellows.

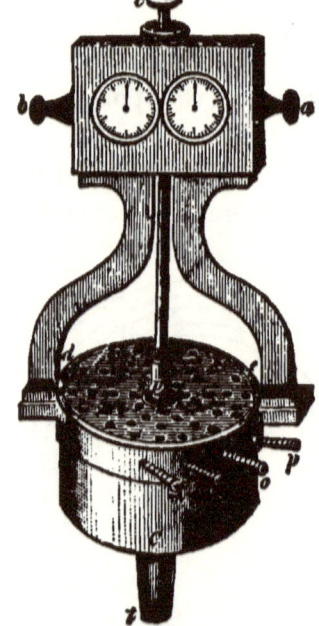

FIG. 191.

Second Method. — Compute the velocity of sound in metres by the formula $v = 333 \ (1 + .0037 \ t)^{\frac{1}{2}}$ in which t is the temperature of the room in centigrade degrees. Now find by trial the length of the shortest air-column giving the loudest resonance when the sounding body is held over it, correcting it for the diameter of the tube. Then $n = \dfrac{v}{l}$ in which l is the wave-length as given by the air-column.

473. Exercise. — Determine the difference of rate of vibration of two sounding bodies, not quite in unison, by counting the number of beats produced during some convenient unit of time, as 30 seconds, and then computing the number per second.

SOUND. 263

XII. LAWS OF VIBRATING RODS AND STRINGS.

474. Apparatus. — Wooden Rod, Sonometer, Siren, Tuning-Forks, etc.

475. Exercise. — Determine the laws governing vibrating rods.

Take a pine strip 2 m. long, 3 cm. wide, and 6 mm. thick, a second one of the same length and width, but 12 mm. thick, and a third one of the same length and thickness, but 6 cm. wide. Clamp these at one end over the edge of a table with small iron clamps, set each in vibration, in succession, and count the number of vibrations made in some convenient period of time. What effect does width and thickness have on the vibration-rate? Now clamp the first one at the middle so that one-half of it is free to vibrate, and determine its rate. What effect does length have on the vibration-rate?

Express as laws the results of these experiments.

476. Exercise. — Determine the laws governing vibrating strings.

A **Sonometer** (Fig. 192) and a **Siren** (Fig. 190) will be required for this experiment. The former is a wooden box about 150 cm. long, 15 cm. wide, and 10 cm. deep. The material is pine, free from pitch, straight-grained, and about 1 cm. thick, except the top, which may be 5 mm. thick; the ends should be of hard wood, and have a thickness of about 25 mm. Near the ends glue triangular pieces of hard wood, to serve as bridges across which to stretch the wires. The depth of these bridges should be about 2 cm., and the distance between them 120 cm. In one of the head pieces insert two stout pins, and in the other two screws, such as are used in pianos, for changing the tension of the wires. In the same end screw a small iron pulley so that one wire may be drawn over it and the tension measured by weights, or by a spring balance. Suitable weights are easily made out of sheet-lead, riveting a number of thicknesses together till the required thickness is obtained. A sliding weight, on a lever, having its fulcrum at the end of the box, might be substituted for the pulley and weights. Beneath one of the wires graduate a scale to half-centimetres.

Now stretch a piano-wire or a catgut string on the sonometer, and produce with the siren, using the inner row of holes, a tone in unison with that given by this string. Without chang-

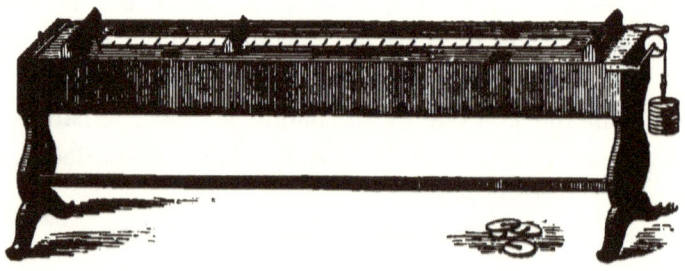

FIG. 192.

ing the speed of the siren-disk compare the tone produced on using the outer circle of holes with that given by the string when its length is reduced one-half by means of a movable bridge placed beneath the middle of the wire. Also compare the tone given by the next to the outer row of holes with that given by the string when its length is but two-thirds of the whole. Also, compare the tone given by the next to the inner row of holes with that given by the string when its length is made four-fifths of the whole. How do these siren-tones compare in vibration-rate? What law must then connect pitch and length of string?

Stretch two piano-wires or catgut strings of the same diameter on the sonometer, one of them in such a way that the tension can be measured. Adjust the tension till the strings sound in unison; there will then be no audible beats. Now, increase the tension of one of the wires till it is in unison with the other one, reduced in length one-half by means of the movable bridge. How have you affected the vibration-rate in reducing the length one-half? What effect must tension have on the rate?

Stretch two wires or strings on the sonometer, the ratio of their weights being 1 : 4. Steel wires No. 27 and 21 B. W. G., or No. 24 and 19, will very nearly fulfil this condition (see Appendix C). Note the tension on the heavy string when its tone is of the same pitch as given by the lighter one. Without changing the tension in either case substitute for the heavy one a wire exactly like the small one. Place a bridge under the first wire at a point where it will cause it to vibrate in unison with the second one. What must have been the difference of pitch of the first two wires? What effect must weight have on pitch?

Stretch with equal tensions a steel and a brass wire of the same gauge-number. Compare the pitch of the sounds. By means of the movable bridge determine the difference in pitch, and hence the ratio of the densities of the two substances.

What three laws for vibrating strings do you derive from the above experiments?

477. Exercise. — Set a large tuning-fork firmly in a block of wood, and clamp it to the table. Tie a metre of fine thread to the end of one of its prongs, pass this thread through a small ring held in a clamp, and to the free end attach a small paper tray of known weight. The thread and the prongs of the fork must lie in the same plane and the thread should have a horizontal position. Now put known weights into the tray till such a tension is secured that, on setting the fork in vibration, the thread will also be thrown into vibration, swinging in one segment. Without changing the length of the thread make the tension one-fourth of what it was at first, and record the number of segments in which the thread vibrates. Try a tension one-ninth of the first. What relation is shown by these results to exist between tension and vibration-rate of strings?

Ascertain the effect on the length of string required, of lowering the pitch of the fork, by sticking a heavy coin to the free end of one of its prongs. What law is here suggested in regard to the relation existing between the vibration-rate and the length of the string?

Substitute for the single thread, one, a third of which is four of these threads twisted together, and the remaining part is single. Change the tension till the string divides into two segments, and observe their relative lengths. What effect does the weight of the string have on the vibration-rate?

A heavy wire stretched on the sonometer may be substituted for the fork. The thread is attached to the centre of the wire and stretched at right angles to the wire. The sonometer wire is set in vibration with a violoncello-bow.

XIII. OVERTONES.

478. Apparatus. —Sonometer, C'-Fork, etc.

479. Exercise. — Tune one of the sonometer strings till it vibrates in unison with a C'-fork. Place the movable bridge at the middle of the scale, that is, at 60, if the scale is 120 cm.

long. The string will now give as its fundamental the first overtone of C'. Place the bridge successively at 40, 30, 24, 20, 17½, 15, 13⅓, and 12. The shorter part of the string in each case will give respectively the second, third, etc., overtones of C'. Ascertain the names of these tones on the Diatonic Scale.

480. Exercise. — Pluck the sonometer string at a point other than its centre, and immediately touch the tip of the finger to the middle of the wire. The fundamental tone ceases, but whence the sound still heard? Compare its pitch with that of the sound emitted by the wire with the bridge at the middle. Ascertain whether the same tone is obtained or not by dampening the wire at the middle, after setting it in vibration by plucking it at the middle. Compare the quality of the tone given by the whole wire when plucked at the middle with that when plucked at some other point. Account for the difference, basing the explanation on the facts revealed by the above experiments. How can any overtone be eliminated from the sound produced by a vibrating string?

In like manner search for other overtones.

481. Exercise. — Tune two strings in unison on the sonometer. Place a bridge under the middle of one of them, and place on the same wire at various points small V-shaped pieces of paper, called *Riders*. Now pluck the whole string at its middle and compare the effect on the riders with that when plucked at some other point. Inference.

Try the bridge at one-third from the end, one-fourth, etc., and ascertain if the whole string vibrates in thirds and fourths when plucked at a point intermediate between thirds and fourths of the string.

XIV. LAWS OF SOUNDING AIR-COLUMNS.

482. Apparatus. — Glass Tubing, Organ-Pipe, etc.

483. Exercise.— Prepare, as described in Art. 456, eight glass tubes, with paper disks and rubber connectors.

1st. Join two of the tubes together; close one end and find by trial a tuning-fork whose tone it enforces. Compare the rate of this fork with that of the fork used in preparing the tubes. Prepare closed tubes that will enforce the tone of such other forks as you have. What do you find that the pitch depends upon?

2d. Prepare open tubes that will enforce the sound produced by the forks employed in the first case. Compare their lengths with those of the corresponding closed tubes. Formulate a law expressing these results.

3d. Connect two closed tubes which enforce the same fork by their closed ends, and select a fork which they now enforce. Remove the partitions, and see whether the air-column will still enforce the same fork or not. What does this show to be the condition of the air at the middle of an open tube which enforces a certain sound?

484. Exercise. — Fit to a glass tube, 2 cm. wide and 40 cm. long, a cork piston. Now set the piston so that the air-column reënforces C'-fork. Blow gently across the open end of the tube, and compare the pitch of the sound with that of the fork. A piece of rubber tubing, flattened at the end by compressing it between the thumb and finger, may be used for the purpose; or, better still, a piece of thin brass tubing flattened at the end, leaving an opening 1 mm. in width. Measure the length of the column. In like manner find the lengths for D',

F′, etc., C″. Compare the ratios of these lengths with those of the vibration-ratios, and write the law suggested.

485. Exercise. — Ascertain the condition of the air in a sounding air-column.

A small **Organ-pipe,** made either of glass, or having one glass side (Fig. 193), and a small wire ring covered with thin paper, will be needed. Sift a little sand over the paper membrane and let it down by a cord into the tube, supported vertically. Blow gently through the mouth-piece, producing the fundamental tone; at the same time move the ring up and down within the tube. Nodes will be known by the sand remaining almost at rest.

Close the tube with a cork having a hole through it for the cord supporting the ring, and again search for nodes.

If the force of the air-current is increased, the pipe will give an overtone. Again search for nodes.

Make a cork piston and slide it along the tube till it reaches the place where a node was found, and compare the sound with that before the piston was introduced. Inference.

Fig. 193.

486. Exercise. — Employing the apparatus of Art. 456, connect several of the tubes together, close one end, and blow gently across the open end to produce the fundamental (lowest) tone.[1] Now increase the force of the blast till an overtone is produced; that is, a sound of higher pitch. Find by trial what

---

[1] This will require considerable skill, as it is difficult to obtain the fundamental tone of a tube of small diameter when the length is great in comparison.

length of closed tube has for its fundamental the previously obtained overtone. Determine from the number of tubes used which overtone you have. In like manner search out other overtones. Repeat the experiment, employing open tubes. What overtones are found to accompany closed tubes ? Open tubes?

XV. HARMONY AND DISCORD.

487. Apparatus. — Tuning-Forks, Sonometer, etc.

488. Exercise. — Compare the pleasantness of the sound produced by two tuning-forks, of the same pitch, with that when to a prong of one of them is cemented a small coin. First, attach the coin near the stem; then at the middle of one branch; and finally at the end. Explain.

If you have the forks, compare C' and D', C' and E', C' and F', C' and G', C' and A', C' and B', C' and C'', etc. Tabulate those which are found to harmonize.

Adjustable tuning-forks are now to be had of dealers in musical wares, at very reasonable prices. With two such forks all these comparisons can be made.

The following substitute may be made for the forks: Set in holes, bored in a board, thirteen test-tubes, each 15 cm. long and about 2 cm. in diameter. Into each pour melted paraffine till an air-column is obtained which responds to some one of the thirteen notes in the octave. Now, proceeding as in Art. 484, any one of these may be sounded, and by using a tube with two branches, any two may be sounded together.

489. Exercise. — Tune two strings on a sonometer to the pitch of C'. By means of the movable bridge under one of the strings produce successively the notes of the musical scale. Determine which harmonize with C', and which do not. Compare other tones by placing a bridge under each wire.

As placing a bridge under a wire will increase the tension, where both ends of the wire are connected to posts, some allowance will have to be made for this.

XVI. VIBRATING PLATES AND BELLS.

490. Apparatus. — Chladni-Plates, Large Bell, etc.

491. Exercise. — Cut out of sheet-brass, 1 mm. thick, a plate 25 cm. square, a triangular one, 25 cm. on each side, and a circular one 25 cm. in diameter. Mount each one on a support, as shown in Fig. 194. Round the edges with a file. Clamp each stand firmly to the table, sift some fine sand evenly over the plate; then, touching the plates at some point with the finger, draw a well-rosined violin-bow across an edge. Try touching the plate at different points. Copy in your note-book the figures formed. Do you discover any relation between the pitch and the figures?

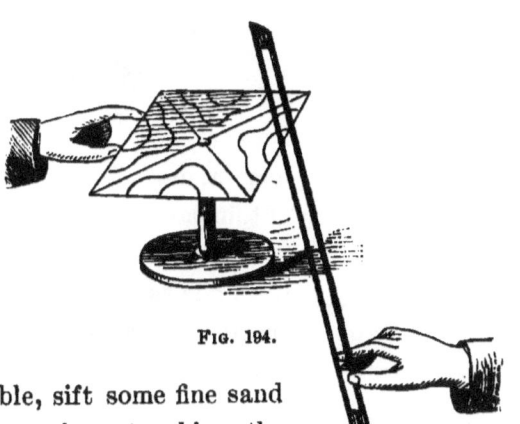

FIG. 194.

492. Exercise. — Using the square plate of the last experiment, make it vibrate so that the only nodal lines are diagonals of the square. While the plate is vibrating, hold over opposite segments or quarters, and close to the plate, stiff paper triangles, having the shape and size of these segments. Observe the effect on the loudness of the sound. Explain.

493. Exercise. — Procure, of a tinsmith, a tube 6 cm. wide and 40 cm. long, having two branches 4 cm. wide, 20 cm. long, and 15 cm. apart at the free end (Fig. 195). Hold the apparatus above the square plate of Art. 491, with the branches within 1 cm. of its surface. Paste a membrane of thin paper across the upper end, and scatter on it a little fine sand. Cause the plate to vibrate in quarters. Compare the effect on the sand when the arms are over a and b (Fig. 196) with that when over a and c.

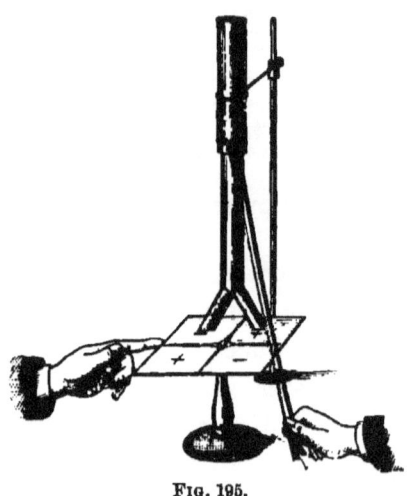

FIG. 195.

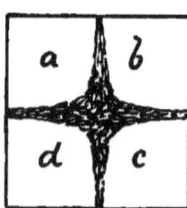

FIG. 196.

494. Exercise. — Invert a large bell, and nearly fill it with water. Scatter lycopodium powder or cork-dust on the surface. Set the bell in vibration by a blow from a small mallet. Determine from the effect on the powder the position and number of the nodal points. A large glass goblet or bell-jar might be substituted for the bell.

495. Exercise. — Employing the bell of the last experiment, determine the position and the number of the nodal points by means of a small glass or ivory ball, suspended by a fine thread. The nodes will be recognized as places where the ball is repelled the least by the vibrating surface.

XVII. ATTRACTION OF VIBRATING BODIES.

496. Apparatus. — Tuning-Fork, Rubber Balloon, etc.

497. Exercise. — Float an inflated rubber balloon on a vessel of water, shielding it carefully from air-currents. Hold near the balloon a large, vibrating tuning-fork, and observe the action of the ball. Repeat several times to make it certain that the effect observed is not accidental. What must be the atmospheric condition about the fork to produce the effect noticed?

498. Exercise. — Sprinkle evenly over the square plate of Art. 491 a mixture of fine sand and lycopodium powder. Cause the plate to vibrate, and compare the effects produced on the powder and sand respectively. Explain.

XVIII. GRAPHIC AND OPTICAL STUDY OF SOUND.

499. Apparatus. — Tuning-Forks, Glass Plates, Vibrograph, Logograph, Manometric Flame Apparatus, Lissajou Apparatus, Sand Pendulum, etc.

500. Exercise. — Cut a wooden block of the form shown in Fig. 197, in which is firmly set the stem of a tuning-fork. Make out of thin brass a narrow rectangular ferrule, with a short pointed arm projecting, to slip over the end of one branch of the fork to serve as a marker. Beneath the point support a pane of glass sufficiently inclined to cause a small plate of glass placed upon it to slide down by gravity. Raise the supporting surface till the style on the fork just touches the glass plate as it slides under it. Smoke the upper surface of the sliding glass, set the fork in vibration by a blow from a small cork mallet, and observe the tracing produced by the style in the smoke as the glass glides down the plane. Guides made of glass tubes cemented to the plane will keep the moving glass from turning from a straight-line course.

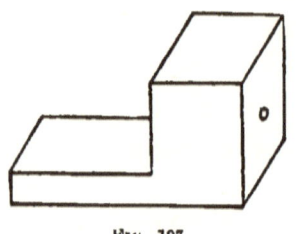

Fig. 197.

Cement a coin to the outer end of the fork, and note the effect on the pitch as shown by the tracing on the glass.

Compare the tracings given by two forks differing an octave in pitch.

In Fig. 198 is shown another device for studying the vibrations of tuning-forks. It consists of a wooden cylinder, A,

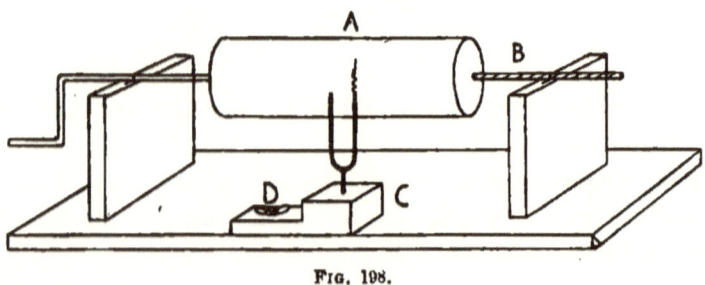

Fig. 198.

mounted on an axis, one end of which, B, has a thread upon it working under the small wire staple which keeps it in place

SOUND.   275

on the standard. The fork is secured by its stem to the block, C, clamped to the base by a screw, D. By this means the fork can be set as close to A as may be desired. The cylinder is smoothly covered with smoked paper, on which the fork-style traces its record as the cylinder is turned by the crank. By measuring the rate of rotation of A, and counting the number of sinuosities made by the fork, the rate of the fork is easily ascertained. By using a second block, two forks may be compared.

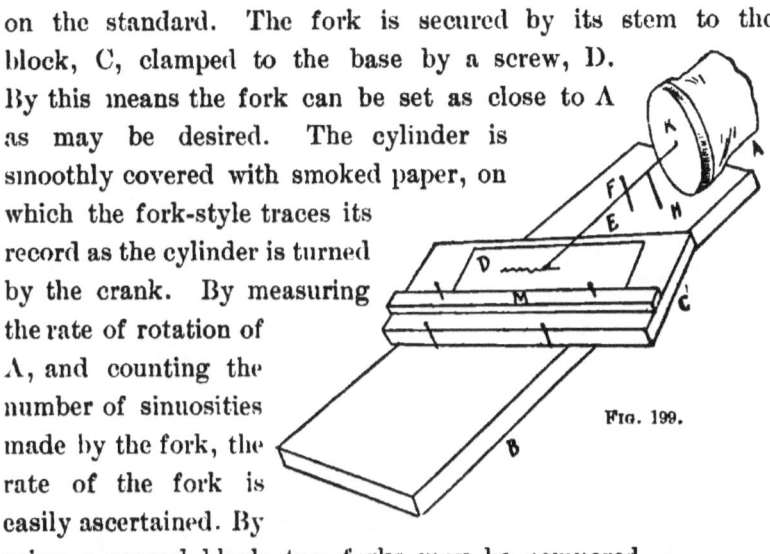

FIG. 199.

501. Exercise. — To study sounds produced by wind instruments, an apparatus constructed as follows will be found quite efficient: Get a good mechanic to turn a wooden mouthpiece like that of a telephone, but somewhat deeper and larger, shown in section in Fig. 200. Glue a parchment-paper diaphragm, K, across it, fastening to its centre, by wax, a slender needle with its point curved down. Fasten this mouth-piece to a board, B, supported in an inclined position. C is a crosspiece to support the smoked glass, D. M is a guide for the glass. C is given such an inclination that the glass plate moves along by gravity. By changing M, the part of the plate under the marker is changed. A stout pin, F, keeps the needle from being dragged out of position, and an elastic, H, drawn down through a hole in B, holds the

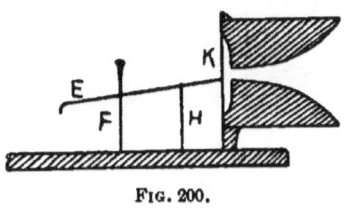

FIG. 200.

276   *PRACTICAL PHYSICS.*

needle in contact with the glass plate. Any sounds produced in front of and close to A will move the needle E, and cause a characteristic line to be traced on D.

502. Exercise. — Cut out of a board 2.5 cm. thick, a strip 4 cm. wide and 25 cm. long. Near one end bore a hole

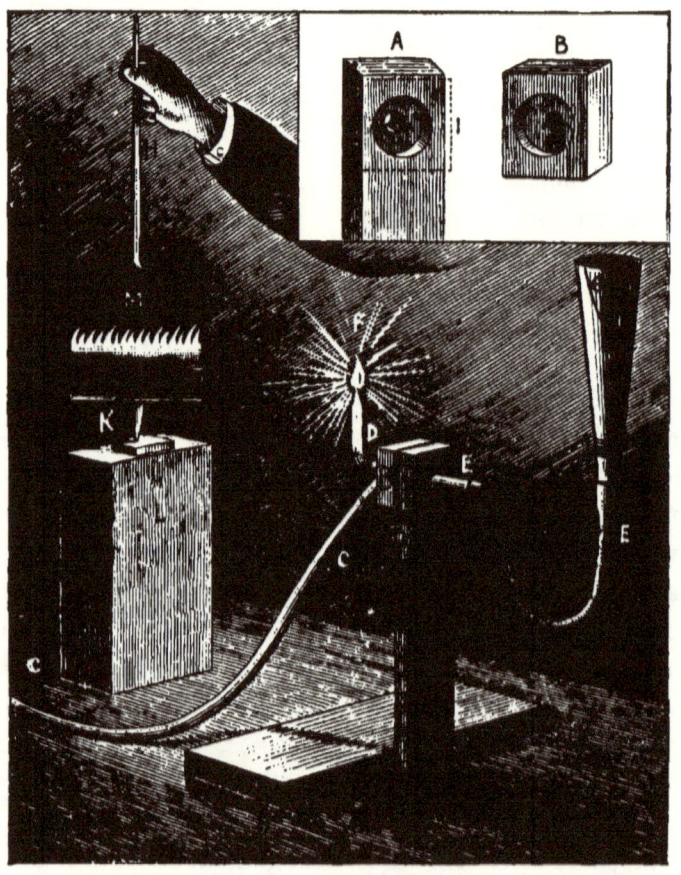

FIG. 201.

2.5 cm. in diameter, half-way through it. Cut a second piece of the same width and 5 cm. long, and bore into it a similar hole. Glue these two pieces together so that the two holes face each other, separating them, however, by a membrane of thin rubber, thus forming a box divided into two parts by this membrane. Nail the long strip to a piece of board for a support. Bore into the strip A, Fig. 201, a small hole leading into the box, and cement into it a short glass tube, E. In B bore two such holes, and cement in the upper one a jet-tube, and in the lower one a short, straight tube. Connect to E a piece of rubber tubing, with a funnel for a mouth-piece. Connect C to the gas-supply. On turning on the gas, a jet can be obtained at F. Cut a cube out of pine, 10 cm. on each edge, as M. Insert in the centres of two opposite faces stout wires, as K and H. Tie on the lateral faces squares of looking-glass. Rotate this apparatus slowly in front of F, and observe the image. Compare it with that when you sing in the funnel mouth-piece, *oo*, in pool. Ascertain the effect of changing the pitch.

Sing the vowel *o*, giving it the pitch of the note B. Sing the vowel *a*, giving it the pitch of F. Sing the vowel *a*, giving it the pitch of C. Try other sounds. Draw a diagram of the flame-figure in each case.

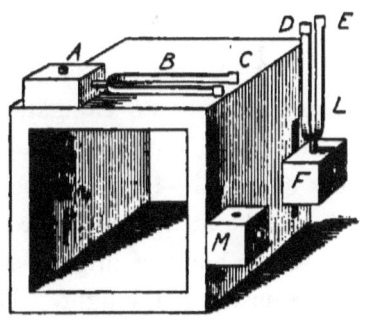

Fig. 202.

503. Exercise.—A complete apparatus for the optical exhibition of vibratory motions is shown in Fig. 202. It consists of a heavy cubical box on two adjacent sides of which are secured, by means of bolts, the blocks A and F, holding the tuning-forks B and L. Slots cut in the frame make it possible to give the forks any desired relative position. Small pieces of mirror, equal in size, are cemented to the lateral faces of the prongs at their outer ends. Flexible steel bars might be used in place of forks, fastening them to the blocks by means of screws. The time of vibration of the forks can be changed by fastening small weights to the prongs with wax.

To use this apparatus, place, a few metres distant from the mirror on the fork at E, a lamp surrounded by a dark chimney, in which is a small hole, giving a small line of light, incident on the mirror. Place the eye so as to see the image in this mirror of the hole in the chimney, and set the fork in vibration. The

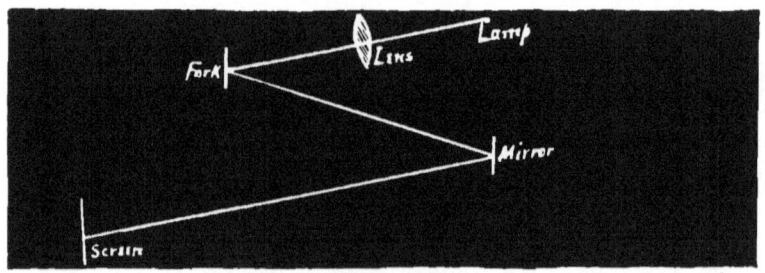

FIG. 203.

image becomes an elongated line, and by rotating the fork a sinuous line is produced, varying in amplitude as that of the fork changes. By receiving the reflected ray on a stationary mirror, the image can be projected on a screen, the distance of the screen and the brightness of the effects being governed by the intensity of the light. Clear definition can be secured by the aid of a lens. Fig. 203 exhibits the order of the apparatus.

Two vibratory motions can be combined by placing the lamp so that the beam of light is incident on the mirror on the fork at D, by which it is reflected to the mirror C, on a second fork, passing thence through a lens to the screen. By moving the

SOUND.

block A to the position M, the forks can be made to vibrate in the same plane or in parallel planes. If the forks are in unison the luminous line on the screen will lengthen and shorten at regular intervals, but if they vary in pitch the beats resulting from the imperfect harmony will be heard, and at the same time a curious pulsating character will distinguish the image. Using the forks as placed in the figure, the combination of the two rectangular vibratory motions is effected, the image obtained depending on the interval of the forks, becoming more and more complex as the ratio of their vibration-numbers becomes less simple.

In Fig. 204 is illustrated a cheap substitute for the foregoing apparatus, the forks being replaced by wooden bars. The credit of the device is due to Mr. George M. Hopkins.

Select a box about 60 cm. square, two flat springs of wood 3 cm. wide, 4 mm. thick, and 60 cm. long, or metal springs 2.5 cm. wide, 1.5 mm. thick, and of the same length as the wooden ones. Secure these at opposite corners by screws to blocks 2.5 cm. thick. Cement to the free ends of each, and in the plane of vibration, a piece of thin cardboard, having a slit 2 mm. wide, parallel with the spring. These cards are placed as near each other as possible. On one of the springs is a sliding-weight of lead, with a set-screw, so that the time of vibration can be changed when necessary. If these springs are now set in vibration, the operator on look-

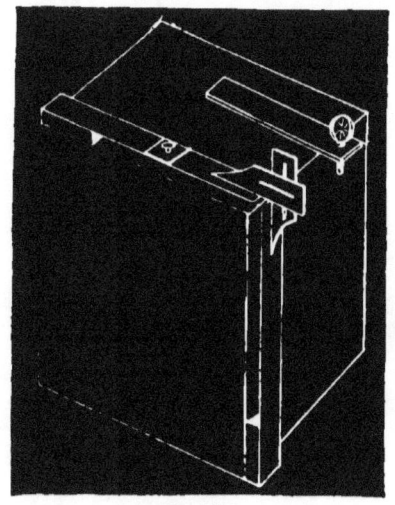

Fig. 204.

ing at the lamp through the two slits will see a band of light, the form depending on the relative rates of the two springs.

By placing a lamp on one side of the slotted cards, and a lens on the other, an image can be projected on the screen. The figures obtained are known as Lissajou's curves. In the next article the same figures are produced by means of Blackburn's pendulum.

504. Exercise. — Construct a frame like the one shown in Fig. 205. The height should be a little over 1 m. The pendulum-bob is supported by a double cord passing through the frame, and winding round a violin-key inserted in the centre of the crosspiece, so that, by turning the key, the bob can be raised or lowered. Sliding on the cord is a thick leather button, r. The bob may be either a small glass funnel, set into a heavy lead disk, or, what is better, a lead disk having a diameter of 8 cm. and a depth of 4 cm., the interior being funnel-shaped, with a circular opening in the bottom 1 mm. in diameter. Fine sand placed in the bob will run out

Fig. 205.

as the bob swings, and will mark out on a board placed beneath it the path described.

Set the leather slide 250 mm. above the centre of gravity of the bob, turn the violin-pin till the whole pendulum is 1 m. long, and swings about 5 mm. above the paper; fill the dish with sand, draw it to one side over a line making an angle of 45° at the centre with that joining the supports. Try in succession the following positions for the slide: 444.4 mm., 562.5 mm., 640 mm., 694.4 mm., 734.6 mm., 765.6 mm., 790.1 mm., the length of the pendulum, as a whole, being kept at 1 m. If the above lengths are not accurately secured, the pendulum will fail to retrace the first-formed curve as it continues vibrating. Draw diagrams of the figures obtained.

An examination of the apparatus will reveal the fact that it consists of two pendulums, one of them 1 m. long, swinging in a plane at right angles to the supporting bar, the other a shorter pendulum, with its point of suspension at the leather slide, swinging in a plane forming any angle at pleasure with the plane of vibration of the longer pendulum. It will also be noticed that the ratios of the lengths of the pendulums in the different cases are $\frac{1}{4}$, $\frac{4}{9}$, $\frac{9}{16}$, $\frac{16}{25}$, $\frac{25}{36}$, $\frac{36}{49}$, $\frac{49}{64}$, $\frac{64}{81}$, respectively. What must be the ratios of their times of vibration? The curve obtained in each case is from the composition of two simple harmonic motions. The pendulum returns to the starting-point after a definite number of oscillations, when the periods are commensurable. When the periods are incommensurable the pendulum never returns to the starting-point, and consequently the curve is not retraced.

XIX. VOCAL ORGANS.

505. Apparatus.— Glass Tube, Sheet Rubber, etc.

506. Exercise.— Roll around the end of a glass tube a strip of thin india-rubber, leaving about 2 cm. of it projecting beyond the tube. Take two opposite portions of the projecting part in the fingers, and stretch it so that a slit is formed. Now blow through the tube, and a sound will be produced. Explain. Ascertain the effect of varying the tension of the rubber. This device is a rough representation of the human larynx.

507. Exercise.— Sing some note of the musical scale, and, without changing the form of the mouth-cavity, hold near it, in a horizontal position, with the prongs over each other, the free ends of a vibrating fork of the same pitch as the note sung. Alter the shape of the mouth-cavity and repeat. Inference.

CHAPTER VII.

LIGHT.

I. SOURCES OF LIGHT.

508. Apparatus. — Porte Lumière Magic Lantern, etc.

509. Exercise. — As parallel rays of considerable intensity are required for many of the experiments with light, below are described several ways of obtaining such a light.

First Method. — Sunlight is brighter than any artificial light, and a horizontal beam can be reflected into the work-room through a south window by means of a **Porte Lumière.** One can be purchased of dealers in physical apparatus for about $5, and can be constructed at even a less outlay, as follows : —

Procure a piece of pine board as long as the width of the window, and about 60 cm. wide. In it cut a round hole 12 cm. in diameter, the centre of which should be the centre of the line parallel to the edge of the board, and 20 cm. from the bottom. On one face of the board, on opposite sides of

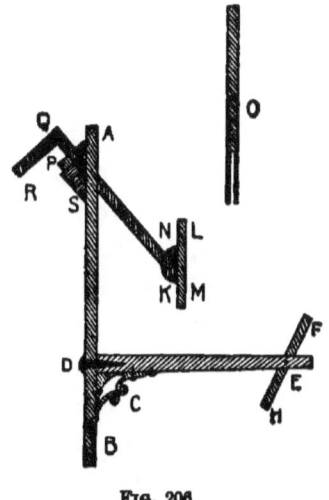

Fig. 206.

the centre of the hole T, and having their inner faces 15 cm. apart, fasten two wooden arms, DE, 32 cm. long, supported by iron braces, C. Figs. 206, 207. Pivot, by screws,

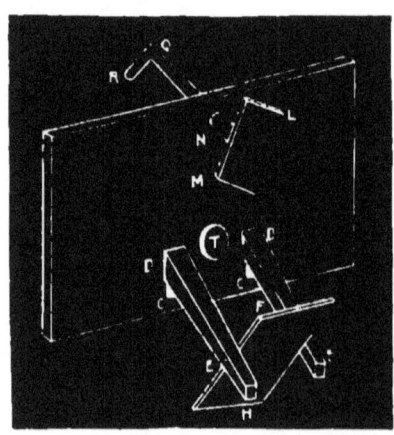

FIG. 207.

between their outer ends, a rectangular board, FH, 2 cm. thick and 15 cm. square, making the distance, DE, 31 cm. Cut a half-round piece, K, 5 mm. thick, with a radius of 7 cm., and fasten it by slender screws to the rectangular piece, LM, 25 cm. by 15 cm. QN is a hard-wood cylinder, 2.5 cm. in diameter; it may be cut from an old broom-handle. In one end, as shown at O, cut a slot so that it will clasp the piece K, to which it is to be secured by a bolt, around which it is free to turn, with slight friction. S is a triangular block, the shape of which is determined by the method illustrated in Fig. 208. On a large sheet of paper lay off the lines DE 31 cm., and DA 40 cm., forming a right angle at D, a point corresponding to the centre of the hole T. At E, with a protractor, make the angle DEQ equal to the latitude of the place. Where EQ cuts DA is the centre of the hole to be bored for QN. The upper angle of the piece S is evidently equal to the angle DEQ. Cut

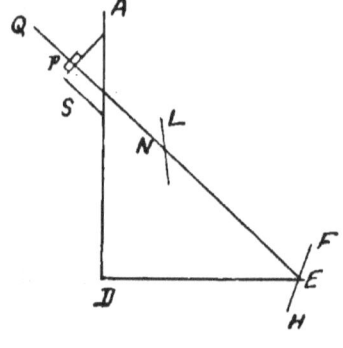

FIG. 208.

this piece S and fasten it to AD, then bore the hole for QN perpendicular to the upper face of S. P is a wooden washer, pinned to QN, to serve as a shoulder, and is so placed that QN extends through the board far enough to bring the lower edge of LM a few centimetres above the top of the hole T. R is a pin at right angles to QN, for convenience in turning it. On the front of both LM and FH fasten a good, plane mirror, completing the apparatus.

Place this porte lumière in a south window, with the board in a vertical position. Determine by trial the inclination of the mirrors so that sunlight incident on LM is reflected to FH, and from it, in a horizontal line, through T. Now, by turning QN on its axis, the beam of light can be kept in one place, notwithstanding the changing position of the sun.

QN is parallel to the polar axis when the porte lumière is in position. The angle between LM and QM changes with the sun's angular distance from the equator, called its declination. If the declination is represented by d, then the angle between ML and QN equals $\frac{90-d}{2}$. It will be found convenient to graduate the piece K to degrees to aid in setting LM.

Second Method. — The best artificial lights are the electric of the arc type and the lime light. For a great many experiments, however, a powerful kerosene-lamp will be found sufficient. With either of these lights, means will have to be adopted to render the rays parallel. To do this, proceed as follows: Place the light in a large wooden box, provided with holes in both the bottom and the top for ventilation. The depth of the box must be such that the heat from the lamp will not set it on fire. In one face of the box, exactly opposite the flame, cut a circular opening, about 10 cm. across, depending on the size of the lens. Tack across this, on the inside, a glass plate to cut off the heat from the lens, which, mounted in a wooden frame, Fig. 209, is tacked over the opening on the outside. This lens had better be plano-convex, and have a focal distance of 15 cm. The light must be situated at its focus, and be incident on its plane surface.

Fig. 209.

Third Method. — The common magic lantern, by removing the objective and adjusting the condensers, can be made to give a very satisfactory beam. This instrument has been so improved of

Fig. 210.

late that in some of its forms it is now capable of giving a light from the use of kerosene varying from 50 to 350 candle-power. Fig. 210 shows one of the simplest in form and best in construction.

II. RECTILINEAR PROPAGATION OF LIGHT.

510. Apparatus. — Cardboard, Tapers, Candles, etc.

511. Exercise.* — Support with retort clamps on a table two small sheets of cardboard in vertical parallel planes. In the centre of one of them cut a hole 2 mm. in diameter (Fig. 211), and place in front of it, at a distance of a few centimetres, a lighted candle or lamp. Compare the images of the source of light formed on the second sheet of cardboard as you vary the relative distances between the three objects. Ascertain the effect of changing the shape of the aperture, and also the size of it.

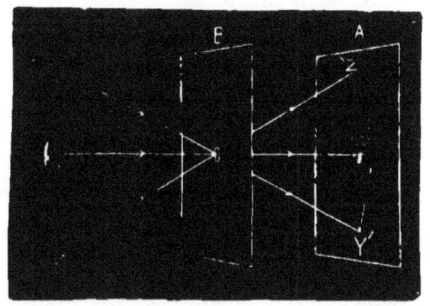

Fig. 211.

Surround the light with a cylinder of cardboard having a small hole 2 mm. in diameter opposite the flame. By changing the position of this cylinder light may be taken from any part of the flame. Now place in front of this opening a sheet of cardboard with an aperture 5 mm. square. Behind this, and distant from it a few centimetres, set the second sheet. Observe the image formed on it and mark its position. Move the cylinder surrounding the light so as to take light from a different part of the flame, and again note the character and the position of the image. In this way take light from such points of the flame as when joined will give an outline of the flame, being careful to move neither the cardboard screens nor the

---

* All those experiments which require a dark room are marked with an asterisk.

light. Now join the centres of the images and compare the figure with that of the flame. Remove the cylinder and note the position of the image. Of what do you find the image to be composed? In what way does the size and shape of the aperture affect the image? Write a full explanation of the whole subject.

512. Exercise.* — Ascertain why an image of every object is not seen imprinted on every other object in accordance with the principle governing the formation of images through small apertures, any point in space being considered as such an aperture.

Set up an apparatus as in the last experiment. Cut a large hole in the cardboard next to the light and cover it with tin foil. Make a pin-hole through the foil and observe the image of the flame formed on the screen. Make other pin-holes and observe the positions of the images. Keep increasing the number till the foil is full of holes and observe the effect. Inference.

513. Exercise.* — Cut out of cardboard two squares 18 cm. and 3 cm. on an edge respectively, and mount them on supports. A simple way would be to secure them with sealing-wax to a heavy strip of cardboard tacked to the edge of a block (Fig. 212). Place these squares successively between a large fan-shaped flame and a screen made of cardboard supported vertically and parallel to the square. Compare the shadows cast by these squares on the screen when the flame is parallel to them. Determine the effect of placing the flame edgewise to the square.

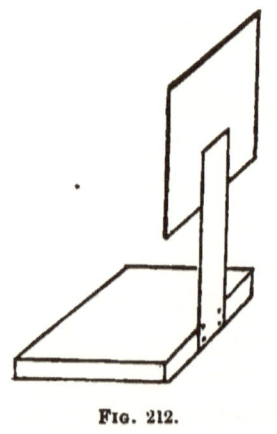

Fig. 212.

Make a pin-hole through the screen in the darkest part of the shadow and look through it toward the flame. Try a hole in the lighter part of the shadow. Try one in the line separating the light from the dark. Try one wholly without the shadow.

Write out a full explanation of shadows based on the facts developed by these experiments.

III. PHOTOMETRY.

514. Apparatus. — Cardboard, Lamps, White Screens, Candles, etc.

515. Exercise.* — Cut from cardboard three squares, 4 cm., 8 cm., and 12 cm., respectively, on an edge, mounting them as directed in Art. 513, making the length of support such that their centres are each at the same distance above the table as the light-emitting point to be used in the experiment. Place around the lamp a paper cylinder, and in it cut a square hole opposite the centre of the flame, but smaller than its least dimensions. Now set the largest square facing the aperture and distant from it about a metre. Between it and the light stand the medium-sized one, finding by trial the place where it just cuts off all the light from the largest. In like manner place the smallest one with reference to the intermediate. Measure the distance of each screen from the light. Compare the total amount of light incident upon the smallest screen with that incident on the intermediate when the first is removed, and also with that incident on the largest when the other two are removed. Compare the amounts on equal areas. What effect do you find distance to have on intensity of light, that is, the degree of illumination of equal areas?

516. Exercise.* — Drop a little hot paraffine or beeswax on a sheet of unsized white paper, and warm the paper till the wax has thoroughly soaked into it. Give the spot a diameter of about 3 cm. Cut from this paper a circular piece 10 cm. in diameter, with the spot in the middle, and cement it to a wire frame provided with a suitable stem for supporting it in a vertical plane. Lay on the table a smooth board 30 cm. wide and 3.5 m. long; draw lengthwise of it a straight line, and graduate it to centimetres. In a rectangular block bore a hole large enough to admit a candle, and in a second block bore four such holes in line and 1 cm. apart. Lengthwise of each of these blocks draw a straight line, bisecting the holes, to be used in determining the position of the candles on the scale on the board. Now support the paper disk on a line between the candles, finding by trial a position for it where the wax spot is either invisible or is least conspicuous to one viewing it from a position in line with the board. Read off on the scale the distance of each set of candles from the disk. Take the mean of a large number of trials. As the wax spot is invisible when the two surfaces of the paper disk are equally illuminated, what relation do you find existing between illuminating power and distance?

517. Exercise.* — Measure the candle-power of a light. Support a white paper screen in a vertical plane. About

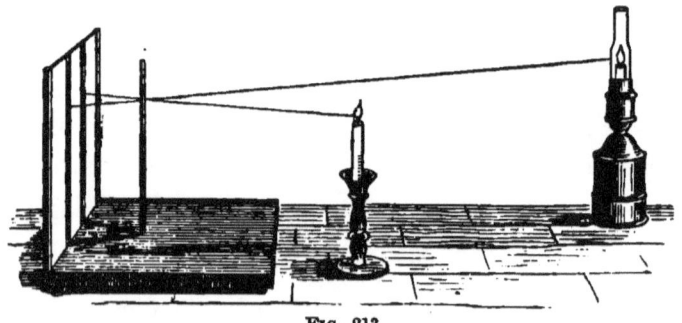

FIG. 213.

10 cm. in front of it fix in a vertical line a wooden rod 1 cm. in diameter, by the aid of wax (Fig. 213). On the same side,

about 30 cm. from the screen, place a lighted sperm candle of the size known as "sixes." Now place the light to be measured at such a distance that the two shadows of the rod seen on the screen are of equal darkness. Then the square of the ratio of the distance of this light from the screen to that of the candle is the candle-power of that light. Why? Make several determinations of the candle-power and find their mean. Tabulate the results.

518. Exercise.* — Another method of determining the candle-power of a light: In Art. 516 substitute for the group of four candles the light whose candle-power is sought. Now move the disk till the waxed spot is dimmest, viewed from either side, and the ratio of the squares of the distances of the lights from the disk will be the candle-power.

IV. REFLECTION OF LIGHT.

a. Regular Reflection.

519. Apparatus. — Porte Lumière, Cardboard, Protractor, Glass Mirror-Plate, Prism, Pane of Glass, Glass Jar, Candle, etc.

520. Exercise. — Cut six blocks of wood, each 4 by 6 by 8 cm., and two pieces of board 8 by 4 by .3 cm. Tack a piece of cardboard of the size of a postal card to each of two of these blocks, using the thin wooden strips as a backing. With a large darning-needle make a hole in each card 15 mm. from the end, removing with a sharp knife the roughness produced. Place the blocks as shown in Fig. 214. On the centre block place a small piece of mirror, and on the outside block stand a lamp or candle with its flame close to the hole in the card. With one eye at the hole in the other card, mark with a

Fig. 214.

long needle, fixing it in place with soft wax, where the line from the image of the light pierces the mirror. Now
draw on a sheet of paper a straight line whose length is the distance between the vertical cards. Erect at its extremities perpendiculars equal to the height of the holes above the upper surface of the blocks. Lay off on the line the distance of the marked point on the mirror from one of the cards, and join the point with the extremities of the perpendiculars. Erect a perpendicular at this point, and measure with a protractor the angles between it and the oblique lines. Make several trials of locating the point on the mirror, and hence get several measurements of these angles. Inference.

521. Exercise. — Determine the law for the reflection of light.

First Method. — On one end of a smooth board, 32 cm. wide by 60 cm. long, draw a circle 30 cm. in diameter, and divide it into degrees by means of a protractor. Pivot at the centre two wooden arms (Fig. 215). Glue to the top one, and at right angles to it, a strip of thin board 2 cm. by 3 cm., with one surface blackened and exactly in line with the axis on which the arms turn. Cement to this vertical strip a piece of mirror through the amalgam of which is cut a narrow line; this line when the mirror is in place must be exactly in the axis. To the outer end of the under arm, or pointer, fasten a needle in a normal to the plane of the graduated circle. To the end of the board farthest from the graduated circle tack a rectangular piece of cardboard, through which, at a point about 1.5 cm. above the board, is a hole 1 mm. in diameter. The free ends of the arms should be bevelled, and a line drawn on this bevelled portion, which, if continued, would pass through the axis. Now place the hole in the cardboard, the line of the mirror, and the mark on the outer end of the mirror-arm in such a position that the line on the mirror bisects the image of the hole, and record the reading of the mirror-arm. Then move this arm a few degrees and bring the other arm around so that the image of the vertical needle, as seen through the hole in the cardboard, coincides with the line on the mirror. Compare the angle through which the mirror-arm was moved, with that between the needle-arm and the second position of the mirror-arm, as shown by the circular scale. Obtain the mean of a number of trials made at different parts of the scale. What law expresses the results?

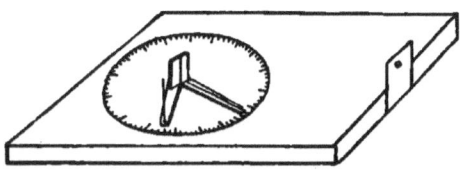

FIG. 215.

Second Method. — On a smooth board about 1 m. by 30 cm. draw a straight line half-way between the ends. Set a rectangular strip of mirror 10 cm. by 20 cm. in the normal plane through this line, fixing it in position by means of a clamp or a little wax. Set in front of this mirror, and distant from it as well as from each other a few centimetres, two darning-needles by sticking them in the board and normal to it, as A and B (Fig. 216). Now set C in line with A and the image of B, and D in line with B and the image of A. Then remove the mirror and draw lines through A and C, and B and D. Measure with a protractor the angles these lines make with the line of the mirror. Vary the position of A and B and obtain the mean of several values for these angles. If each of these angles is subtracted from 90° we have either the angle of incidence or the angle of reflection in each case, as it will be observed that C is in line with A and the images of both B and D, and D is in line with B and the images of both A and C. What relation is here established between these angles?

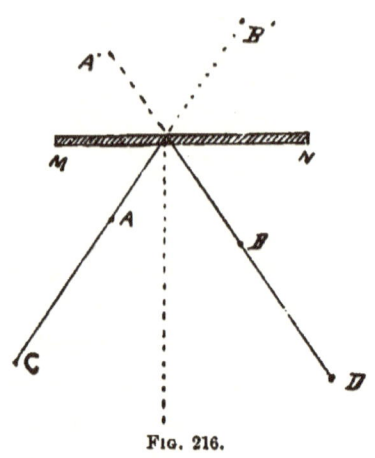

Fig. 216.

To insure greater accuracy the needles which locate lines should not be placed very close together.

522. Exercise. — Measure the angle of a prism.

*First Method.** — Cut a slit 1 mm. wide and 3 cm. long in a square of cardboard, and fasten it across the opening of the lantern, or porte lumière, so that a vertical ribbon of light is obtained. Place the prism with the angle to be measured towards the opening in the cardboard, and approximately bisected by the central plane of the ribbon of light (Fig. 217). Part of the ribbon of light will be incident on one face of the prism and be reflected from it, and the other part will be similarly disposed of by the other face. Now stick perpendicularly in the board supporting the prism two darning-needles, N_1 and N_2, one on each side of the prism and in the reflected beam. Then set N_3 and N_4, two other darning-needles, in the centres of the shadows cast by N_1 and N_2, respectively, and distant several centimetres. Remove the prism, draw lines through the points marked by the needles, and measure with the protractor the angle they form at D. Half of this angle is the angle of the prism, since $EDF = EAF + AED + AFD = EAF + AEH + AFK = 2EAF$.

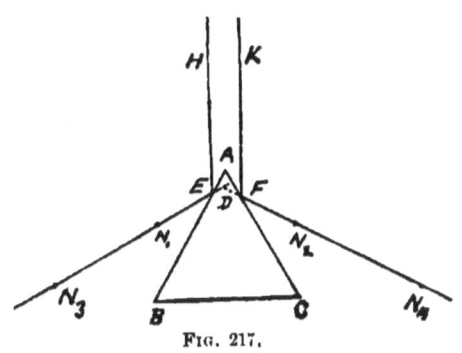

FIG. 217.

FIG. 218.

Second Method. — Place the prism on the stand of the **Spectrometer** (Fig. 218), with the angle to be measured turned toward the slit, the plane of light approximately bisecting the angle. Now move the telescope till the intersection of the cross-wires is exactly on the image of the slit reflected; first from one face, and, secondly, from the other face of the prism. The difference between the readings for these two positions, given by the vernier attached to the telescope-arm on the horizontal scale around the edge of the table of the instrument, will be double the angle of the prism.

It will be readily seen that this method differs from the first one, simply in the use of a telescope and verniered scale to secure more accurate determinations of the angles involved.

b. Diffused Reflection.

523. Exercise.* — Find the effect of an irregular surface on light.

Let a beam of light be incident successively on a good plane mirror, a tarnished surface, a pane of window-glass, a pane of ground-glass, and a sheet of white writing-paper. Compare the brightness of the spots of light formed on the wall by reflection from these surfaces. Inference.

524. Exercise.* — Fill a large glass jar with smoke by burning touch-paper within it. Cover up the top of the jar with a piece of cardboard, through which is a hole 1 cm. in diameter. Set a plane mirror at an angle so that it will reflect a beam of light into the jar through the opening in the cover. Compare the effect with that when an empty jar is used. Explain.

c. Amount of Light Reflected.

525. Exercise.* — Ascertain if the amount of light reflected from a surface is affected by the size of the incident angle.

LIGHT. 297

Support a small plane mirror in a retort-holder, giving it such an inclination that a strong beam of light incident upon it is reflected to the surface of a shallow basin filled with water, striking it obliquely. By varying the distance of the water from the mirror, as well as the inclination of the mirror, any incident angle at the surface of the water may be obtained at pleasure. Compare the brightness of the reflection from the water on the wall as given under different angles. Inference.

526. Exercise.[*] — Place a sheet of writing-paper on a table, and near it a lighted candle. Find a position for the eye where an image of the candle can be seen in the paper, and observe the size of the angles of incidence and of reflection. Inference.

V. MIRRORS.

527. Apparatus. — Plane Mirrors, Darning-Needles, Small Iron Rings, Pane of Glass, Bottle, Candles, Cardboard, Protractor, Glass Tube, Concave and Convex Mirrors, Porte Lumière, Lenses, Small White Screens, Spherometer, Scale, etc.

a. Plane Mirrors.

528. Exercise. — Find where the image of an object in a plane mirror is situated.

First Method.—Select a smooth board, about 1 m. long and 30 cm. wide, draw a line across the middle of it and support a plane mirror in the normal plane containing this line, either by means of a clamp or sealing-wax. Set a darning-needle perpendicular to the board, about 15 cm. in front of the mirror, and also a second one in an exact line with the first one and its image in the mirror (Fig. 219), at, say, 30 cm. from it. As far as possible to one side of these set two more needles, so that the line of sight through them passes exactly through the image of the first needle set. Now remove the mirror and draw lines through the points marked by the needles, producing them till they intersect. Measure the distance of this intersection from the reflecting surface of the mirror, that is, from the line drawn across the board, and also measure the distance of the first needle set from the same point. Make a number of trials and obtain the mean. Inference.

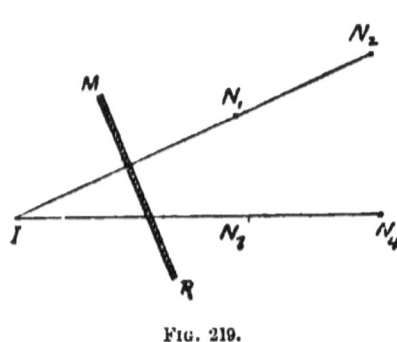

Fig. 219.

Second Method. — Procure at a hardware store two iron rings, about 3 cm. across. Cement across them fine wires, forming diameters at right angles. Draw across the middle of a smooth board, 2 m. long, a straight line, and a second one lengthwise of the board, and perpendicular to the first. Graduate this last line to half-centimetres each way from the intersection. Secure with sealing-wax in a vertical plane and on the zero-line a piece of mirror, 10 cm. square, with the amalgam removed from the upper half (Fig. 220). Cement each ring to the end of a wire, fixed

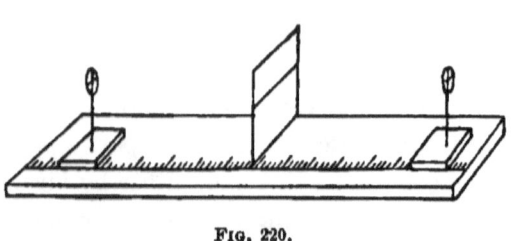

Fig. 220.

vertically in a small block of wood, the length of wire to be such that the centre of each ring is 5 cm. above the board supporting the mirror. Draw on the block in the plane of the ring a line to aid in obtaining the position of the ring on the linear scale. Set one of the rings behind the mirror, distant from it several centimetres, and the other one in front. Now, the upper half of the first ring can be seen through the unsilvered part of the mirror when the eye is placed level with the line dividing the silvered from the unsilvered part, and the lower half of the other one will give an image behind the mirror. Carefully adjust the position of the ring in front of the mirror till the image of its lower half forms an exact continuation of the upper half of the one back of the mirror, making the ring and cross seem complete on moving the eye either to the right or the left. This position is easily determined by observing that when the front ring is too near the mirror, on moving the eye to the right the image and object will appear to separate, the image going to the left; and if too far from the mirror, the reverse is the case. Now read off on the scale the distance of each ring from the mirror. Repeat the experiment several times, and obtain the mean of the measurements. Inference.

529. Exercise.* — Procure a large pane of window-glass, say, 12 by 20. Support it in a vertical plane back of a rectangular opening, smaller than the glass, cut in a large sheet of cardboard. The planes of the cardboard and glass make a small angle with each other (Fig. 221). Place a lighted candle, L, between the screen and the glass, leaving the image alone visible through the opening. Behind the glass set a large bottle of water, finding, by trial, a position where to one standing in front of the opening in the cardboard the candle appears to be burning within the bottle. Compare the distances of the candle and the bottle from the glass plate. Inference.

FIG. 221.

The above experiment, in addition to locating the apparent position of the image of an object in a plane mirror, also illustrates one method of producing spectral phenomena on the stage.

530. Exercise. — Determine the law for the number of images in cases of multiple reflection.

Cut out of mirror-plate two pieces, each 15 cm. square. Hinge them together at one edge by pasting a strip of cotton across the backs of both. Stand them vertically on a sheet of paper on which is drawn a protractor scale, placing the axis of the mirrors in line with the centre of the circle (Fig. 222). Now stand a lighted taper between the mirrors, and count the number of images for any specified angle of the mirrors. Try 90°, 60°, 45°, and 40°. Note the relation sustained by the angle of the mirrors to the entire circle, and compare in each case with the number of images. Account for so many images being visible.

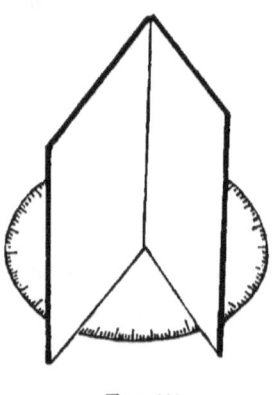

FIG. 222.

Place two such mirrors parallel, and note the number of

images. Explain. Show that the result agrees with the law governing the number of images.

Note the image of a candle in a very thick mirror. Explain. Verify your explanation by blackening one surface of a thick piece of glass, with the amalgam removed, and observing the image of the candle in it. To blacken the glass hold it in the smoke emitted by burning turpentine or camphor.

531. Exercise. — Coat with asphaltum varnish a glass tube 30 cm. long and 2.5 cm. in diameter. Make a pin-hole in a small piece of cardboard; hold it across one end of the tube, with the pin-hole in the axis of the tube. Now, with the open end to the eye, look at the bright sky. Describe the appearance, and account for it.

b. Curved Mirrors.

532. Exercise.* — Determine how a concave mirror disposes of incident light.

First. — Let a small beam of light be incident, perpendicularly on the mirror. The floating dust will mark out the path of the rays, showing that the rays are reflected through a point called the principal focus. Measure the distance from this point to the vertex of the mirror. Twice this distance gives the radius of curvature, and hence enables you to locate the centre. Mark both of these points in front of the mirror with pointers held by clamps, or by wires set in large corks for supports.

Sunlight can be used for finding the principal focus, by ascertaining the point of greatest intensity of light in front of the mirror; that is, where the light, reflected on a piece of cardboard, produces the smallest bright spot.

Secondly. — Let a pencil of light diverging from a point beyond the centre of curvature be incident on the mirror, and find where it focuses with reference to the centre and the principal focus, as shown by the illuminated dust-particles. In like manner locate in succession the focus for a pencil of light diverging from the centre, from a point between the centre and the principal focus, from the principal focus, and from a point between the principal focus and the mirror.

Thirdly. — Let a converging pencil be incident on the mirror, and determine its course after reflection.

A simple way to obtain a diverging pencil for the above experiments is to let a beam of light be incident on a lens having a focal distance of about 50 cm. The lens will refract the light through a point, producing a pencil diverging from its focus. For a converging pencil, the part between the lens and the focus may be used.

533. Exercise.* — Ascertain the properties of a convex spherical mirror.

Proceed as in the last experiment, making such modifications as the problem demands.

534. Exercise.* — Determine the character of the images of an object given by spherical mirrors.

Support the concave mirror in a vertical position on a long table, and place in front of it a lighted candle at a distance greater than the radius of curvature of the mirror. On the same side of the mirror support a small paper screen, finding by trial a position for it where a sharply defined image of the candle is formed on it. Compare the distances of the object and image respectively from the mirror; also, size, position, etc. Now gradually move the candle nearer to the mirror and determine the character of the image for each new position, particularly when the light is at the centre of curvature, between the centre and the principal focus, at the principal focus, and between the focus and the mirror.

In the case of the convex mirror, what is the character of the image? Ascertain the effect that the position of the object has upon the character of the image.

535. Exercise. — Measure the radius of curvature, and hence the focal distance of a concave spherical mirror.

First Method.* — Support the mirror vertically on a long table; place in front of it a sheet of cardboard in which is cut a hole 1 cm. square. Illuminate the opening by placing a lamp back of it, and place a screen to receive the image of it formed by the mirror. Move the screen till the outline of the image is brightest and has the sharpest outline. Measure the distance of the object from the mirror, and also that of the image. If these are represented by a and b respectively, and the radius by R, then

$$\frac{1}{a} + \frac{1}{b} = \frac{2}{R} = \frac{1}{f} \text{ whence } R = \frac{2ab}{a+b} \text{ and } f = \frac{ab}{a+b}.$$

Second Method. — Insert a small piece of stiff white paper in the eye of a darning-needle to serve as an index, and at the same time to make it more conspicuous. Set it vertically in front of the mirror, finding by trial a place where the point of the paper coincides with the image of it seen in mid-air. Measure the distance from the pointer to the vertex of the mirror. Obtain the average of several trials; the result will be the radius of curvature. Why? How would you get the focal distance?

Third Method. — Place the spherometer on the curved surface, and, proceeding as in Art. 11, obtain the altitude of the segment that would be cut off by the plane of the three legs of the instrument. Let h = this altitude, and s = the side of the equilateral triangle formed by the feet. Then $R = \dfrac{s^2}{6h} + \dfrac{h}{2}$, a relation deduced geometrically from the properties of a sphere.

This method cannot be applied to the ordinary glass mirrors, as they are usually plano-convex lenses silvered on the convex surface.

536. Exercise. — Measure the radius of curvature of a convex spherical mirror.

First Method. — Locate the position of the image as directed in Art. 528. Measure the distance of the object, and also the image, from the vertex of the mirror. Then

$$\frac{1}{b} - \frac{1}{a} = \frac{2}{R}, \text{ whence } R = \frac{2ab}{a-b}.$$

Second Method.* — Cut a round hole 5 cm. in diameter in a sheet of cardboard, and support it several centimetres in front of the mirror and perpendicular to its axis (Fig. 223). Let a strong beam of light be incident on the cardboard, part passing through the opening, making a round image on the mirror, and a second one on the cardboard, by reflection.

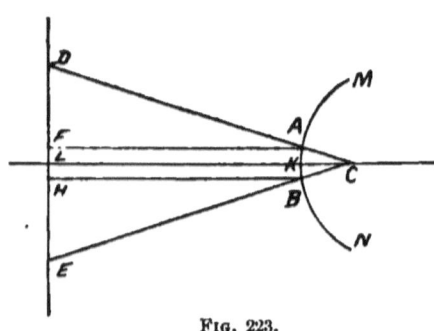

FIG. 223.

Measure the diameter, DE, of the spot on the screen; FH, or AB of the opening or spot on the mirror; and LK, the distance from the screen to the mirror. Then if $AB = a$, $DE = b$, $LK = c$, and $KC = f = \frac{1}{2}R$, we have, by geometry,

$$\frac{a}{b} = \frac{\frac{1}{2}R}{c + \frac{1}{2}R} = \frac{R}{2c + R}, \text{ whence } R = \frac{2ac}{b-a}.$$

Third Method. — Employ the spherometer as in Art. 535.

537. Exercise.* — Compare as to brightness and sharpness of outline the image given by a concave mirror when the light is cut off from all but the central portion, with that when the whole surface of the mirror is used. Bend a piece of bright tin into a semicircular form, place it on a sheet of white paper with its concave surface toward the light, and notice how it reflects the rays. Explain.

VI. SINGLE REFRACTION OF LIGHT.

538. Apparatus. — Porte Lumière, Rectangular Battery-Jar, Protractor, Cardboard, Prisms, Darning-Needles, Spectrometer, Plane Mirrors, Globular Receiver, etc.

539. Exercise.*—Measure the index of refraction of a liquid. Cover one face of a large glass rectangular battery-jar with paper, out of the centre of the covering having cut as large a circle as possible. Across this circular opening draw a horizontal and a vertical diameter (Fig. 224). Graduate the margin of the circle to degrees, the extremities of the vertical diameter to be marked zero, and those of the horizontal one 90°. Provide a strip of cardboard considerably wider and longer than the top of the jar, and cut in it, cross-wise, a slit 2 mm. wide and 5 cm. long.

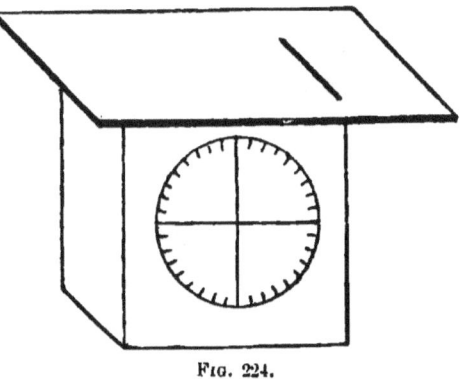

Fig. 224.

Fill the jar with water exactly to the horizontal diameter of the circle. Place the cardboard containing the slit on top of the jar. With a plane mirror supported in a retort-holder reflect a strong beam of light through the slit at such an angle as to be incident on the liquid exactly back of the centre of the circle. Read off on the scale the angle that the incident ray makes with the vertical, and also that made by the refracted ray seen in the liquid. Change the position of the slit and the mirror so as to get a number of these angular measurements for different beams. Divide the sine of each angle of incidence (see Table XX.) by the sine of the corresponding angle of refraction, and the average of these quotients will be the index of refraction.

A large square bottle, or a tin tank with one glass face, may be substituted for the battery-jar. In using the battery-jar it will render the rays more distinct to cement black paper on the faces.

Any of the following liquids may be used for this experiment: Water, alcohol, naphtha, turpentine, kerosene, etc.

540. Exercise. — Determine the index of refraction of glass.

*First Method.** — Tack across the opening of the porte lumière, or lantern, a piece of cardboard in which is cut a vertical slit 1 mm. wide and 3 cm. long. In the path of the ribbon of light which passes through the slit place a glass prism of known angle (see Art. 522), supported on a smooth, horizontal board, with its edges parallel to the plane of the ribbon of light, and intercepting it at such an angle that it suffers the least devia-

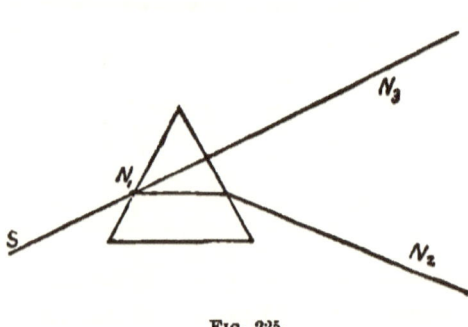

Fig. 225.

tion from its course; a result recognized by slowly turning the prism about its axis till the bright-colored image on the wall departs least from the line of light before it suffers refraction. Now set perpendicularly in the board a darning-needle, N_1, to mark the place where the light is incident on the prism (Fig. 225). Set a second one, N_2, in the shadow cast by N_1. Remove the prism and set a third one, N_3, in the shadow now cast by N_1. Draw lines through the points marked by the needles, and measure with a protractor the angles between them. Make a number of trials, and determine their average. This will be the angle of deviation for the prism. If $a =$ the angle of the prism and $d =$ the deviation, then the index of refraction,

$$\text{or } i = \frac{\sin \tfrac{1}{2}(a+d)}{\sin \tfrac{1}{2}a}.$$

Second Method. — Place the prism on the stand of the spectrometer of Art. 522, a form in which the telescope is attached to a movable arm carrying an index moving over a graduated circle. Place the prism in the position of least deviation. This is found by observing the slit through the telescope, at the same time slowly turning the prism till the dark vertical line seen in the green part of the image moves out of the field in the same direction on turning the prism either way. Now read the index of the telescope-arm when the cross-hairs are exactly on the dark line in the green. If not visible, use the dark line in the yellow. Then remove the prism, and obtain the reading when the slit is in the field of the telescope. Obtain the average of several determinations of these readings, and their difference will be the angle of deviation for the prism. To obtain the index of refraction, apply the formula of the first method.

Obtain the index, using dark lines seen in other colors of the spectrum.

541. Exercise.* — Using the apparatus of Art. 539, place the cardboard against the end, so that the slit is near the bottom of the jar. Now place a plane mirror, so as to throw a beam of light on to a second mirror lying on the table close to the slit, and incident at such an angle that it passes through the slit into the liquid, reaching the surface opposite the centre of the circle. Observe the course of the light after incidence, taking the readings on the graduated scale. Explain.

Raise the slit, changing, accordingly, the position of the mirror till the ray, after incidence, passes out along the surface. Measure the angle of incidence, thus obtaining the critical angle for the liquid. Try different liquids.

542. Exercise.* — Procure a globular receiver, about 10 cm. in diameter, with two tubulures (Fig. 226). Close each neck with good corks, through which pass as large glass tubes as possible. Coat the outside of the globe with black varnish, except the area included within the small circle, indicated by the dotted line, opposite the neck to be used as the horizontal one. Fill the globe with water, having first closed the horizontal tube with a cork; connect the upper one with a rubber tube to a tank supplying water, and adjust the apparatus in the opening of the porte lumière so that the light is incident on the unvarnished part, and all light is cut off from entering the room. Now remove the stopper from the exit-tube, and notice how the light follows the stream of water. Explain. A strip of red glass placed between the globe and the source of light produces a very striking effect.

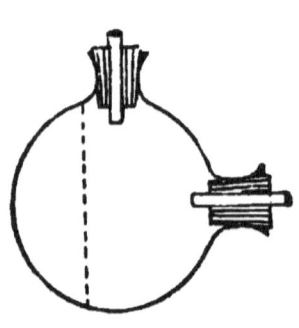

Fig. 226.

LIGHT. 309

543. Exercise.* — Let a small beam of light be incident perpendicularly on one of the faces about the right angle of what is known as the right-angled prism. Observe the course of the light. Explain.

VII. LENSES.

544. Apparatus. — **Porte Lumière, Convex and Concave Lenses, Cardboard, Scales, Telescope, Spherometer, Candles,** etc.

545. Exercise. — Measure the principal focal distance of a convex lens.

First Method. — Tack a piece of cardboard, about 15 cm. square, across one end of a common lath (Fig. 227). Beginning at the cardboard, lay off a half-centimetre scale on the wooden strip. Now hold the lens on the strip parallel to the screen, direct the apparatus toward the sun, and slide the lens along till a position is found where the light forms the smallest and brightest image of the sun on the screen. Read off on the scale the distance of the lens from the screen for the focal distance required. Obtain the mean of several independent determinations.

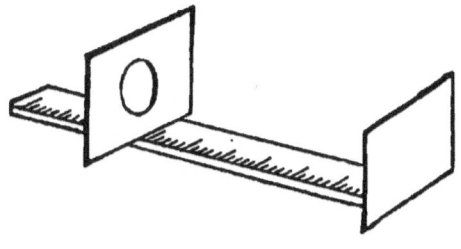

FIG. 227.

*Second Method.** — Support the lens in a vertical plane, placing on one side of it a cardboard screen, and on the other a sheet of cardboard with a circular aperture, 1 cm. in diameter, having two threads, mutually perpendicular, stretched across its centre, and illuminated on the opposite side from the lens by a lamp or candle placed near the opening. Adjust the apparatus so that the centre of the opening lies in the principal axis of the lens. Now find by trial a position for the lens between these sheets of cardboard where a sharply defined image of the threads is formed on the other screen. Measure the distances between the lens and the image and object respectively, representing them by a and b, then the focal distance $= \dfrac{ab}{a+b}$. Obtain the mean of several determinations.

This method neglects the thickness of the lens. As this is usually small, the value of f would not be materially altered.

Third Method.* — Using the apparatus of the last method, adjust the distances so that the image and object are equal in size. Then one-fourth of the distance between them is the focal distance of the lens. Obtain the mean of several determinations.

Fourth Method. — Focus a small telescope, as a common spy-glass, on some object so far distant that the rays from it may be considered as parallel without sensible error. Now place the lens, whose focal distance is to be measured, in front of the telescope, and look through the two at some figure on a piece of cardboard held in a clamp. Move the cardboard to such a distance from the lens as to be clearly visible. The distance of the cardboard from the lens is the focal distance sought. Obtain the mean of several trials.

Fifth Method. — If the lens has sufficient surface, the radius of curvature of each surface can be determined by the spherometer, as in the case of mirrors (Art. 535). Representing by r and r' the radii of curvature of the two faces, and by i, the index of refraction, then $f = \dfrac{1}{i-1}\left(\dfrac{rr'}{r+r'}\right)$. The average in-

dex of crown glass may be taken as 1.5 without sensible error.
Hence, $f = \dfrac{2\,rr'}{r + r'}$. If both surfaces have the same curvature, then $r = r'$ and $f = r$, that is, the focal distance equals the radius of curvature. If the lens is plano-convex, then $r' = \infty$ and $f = 2\,r$. For the meniscus lens the radius of curvature for the concave surface must be taken as negative.

546. Exercise. — Measure the principal focal distance of a concave lens.

First Method.* — Cover the face of the lens, that is turned away from the light, with paper, in which is cut a smooth, round hole 2 cm. in diameter, exactly over the optical centre. Support the lens so that a beam of light parallel to the principal axis is incident upon it. Back of the lens, and parallel to it, place a cardboard screen. The light, after passing through the lens, will diverge as if it came from the principal focus, and will form a round image on the screen.

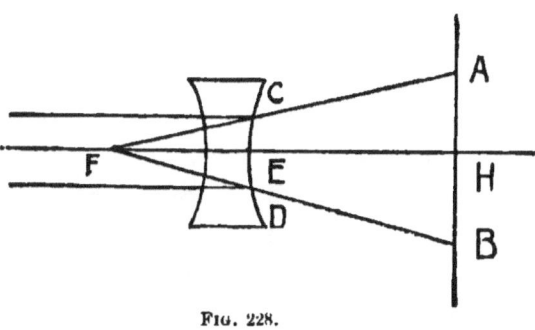

Fig. 228.

Hence, if we measure the distance AB (Fig. 228), the diameter of the image; the distance CD, the diameter of the aperture; and the distance EH from the lens to the screen; we can compute the focal distance EF, neglecting the thickness of the lens, as it is usually small. Representing these distances by a, b, and c, respectively, we derive from the triangles in the figure

$$f = \frac{bc}{a-b}.$$

By moving the screen till $AB = 2CD$, the distance of the screen from the lens is the focal distance.

Try both ways, obtaining the mean of several determinations.

Second Method. — Place in contact with the concave lens a convex one, making the combination equivalent to a convex lens. Measure the focal distance of the combination, as in Art. 545, and also of the convex lens alone. Then $f = \frac{Ff'}{F-f'}$, in which F is the focal distance of the combination, and f' of the convex lens.

Third Method. — Employ the spherometer as in Art. 545, making the r's in the formula negative when the surface is concave.

547. Exercise.* — Ascertain how a lens disposes of a pencil of light.

First. — After determining the focal distance, support the lens in a clamp, and let a diverging pencil of light be incident upon it, as in Art. 532. Ascertain where the rays focus when the light diverges from a point beyond twice the focal distance, at twice the focal distance, at less than twice but beyond the focus, at the focus, and, finally, between the focus and the lens. Test both convex and concave lenses.

Secondly. — Try a converging pencil. In the case of a concave lens vary the degree of convergence by using lenses of different focal distances to produce the pencil of light.

548. Exercise.* — Determine the character of the images formed by lenses.

First. — Place on a long table a cardboard screen, a mounted convex lens of known focal distance, and a lighted lamp or candle, arranging them in the order named, and with their centres in the same horizontal line. Place the light successively at more than twice the focal distance, at twice the focal distance, at less than twice but beyond the focus, and, finally, between the focus and the lens, determining in each case, if possible, the position of the screen where the well-defined image of the flame is formed upon it. Compare the image in each case with the object as regards size and position.

Secondly. — Compare the image formed when a lens of small diameter is used with that when one considerably larger is employed and adjusted to give an image of the same size.

Thirdly. — Cover a lens with a paper disk in which is cut a ring of holes near the circumference, and also a ring near the centre. Mount the lens so that a beam of light is incident upon it and focuses on a screen. Ascertain if the outer ring of holes focuses at the same distance from the lens as the inner row.

Fourthly. — Compare the definition of the image formed when light is admitted through the whole lens with that given when the light incident on the outer edge of the lens is cut off by a paper ring placed over it.

Fifthly. — Study the images given by a concave lens.

549. Exercise. — Measure the magnifying power of a lens.

First Method. — Measure the focal distance of the lens, then the magnifying power $= \dfrac{25\ cm.}{f} + 1$, in which 25 cm. is taken as the distance of distinct vision.

Second Method. — Focus the lens on a finely divided scale securing clear definition. Place a second scale at the distance of distinct vision, 25 cm. Now look with one eye through the lens at the one scale, and with the other eye at the second one. On the eye's becoming accustomed to the conditions of the experiment, the magnified image of one scale will be seen superposed on the other. Count how many divisions of the magnified scale exactly cover a certain number on the other. Divide the latter by the former, the quotient will be the magnifying power.

VIII. DISPERSION.

550. Apparatus. — Porte Lumière, Cardboard, Lenses, Prisms, Glass Bulb, etc.

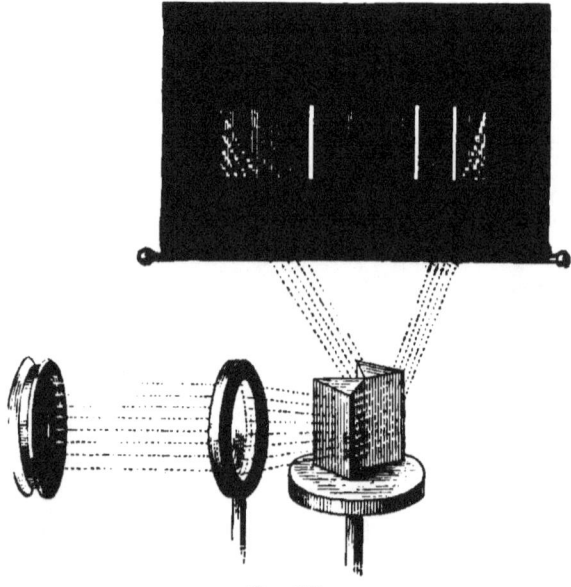

Fig. 229.

551. Exercise.* — Tack across the opening of the porte lumière, or lantern, a piece of cardboard in which is cut a very narrow slit with straight smooth edges. Project with a convex lens of about 30 cm. focus an image of this slit on a distant white screen. Close to the lens, between it and the screen, place a prism of crown glass (Fig. 229), of 60° angle, and change the position of the screen, keeping it at the same distance from the lens, to a position where the light falls perpendicularly upon it. Turn the prism around till the least deviation is secured. Describe the image.

Compare the spectrum given by a prism of crown glass with that given by one of flint glass, or one of carbon disulphide.

Try two prisms, setting the second one behind the first, with its base turned in the same direction.

To make a carbon disulphide prism proceed as follows: —

Cut off the ends of a stout glass tube, 4 cm. in diameter, by planes inclined to each other at an angle of 60°, and their line of intersection perpendicular to the axis of the tube (see page 354). Grind the ends smooth on a flat sandstone. Drill a hole through the side of the tube. This hole had better be made before cutting off the ends, to avoid lost labor from a possible breaking of the tube in drilling it. Cement with good glue two pieces of the best of thin plate-glass on these faces; mirror-plate with the amalgam scratched off is the best, as it is more likely to be true. When dry, fill the angles on the outside between the glass plate and the tube with a cement made of glue dissolved in common molasses by the aid of heat. Now fill the cell through the hole in the side, cork, and cover the cork with cement.

The cell must not be filled in the vicinity of a flame, as carbon disulphide is very inflammable.

552. Exercise.* — Project, as in the last experiment, a spectrum of some strong light on a screen. Cut a hole through this screen so that one color passes through and is incident on a second prism. Ascertain if any new colors are obtained by the second dispersion. Inference.

White printing-paper or cotton cloth, pasted over a light wooden frame, will make a suitable screen.

553. Exercise.* — Project, as directed in Art. 551, a spectrum on a screen, using a short slit. Behind the prism place a second prism with its edges perpendicular to the edges of the first, and in a position to receive the light on issuing from the first. By this arrangement there will be incident on the second prism rays of each color, each one by itself, making it possible to determine whether a prism refracts each color alike or not; for if it does, the new spectrum will occupy a position parallel to the old one. Inference.

554. Exercise.* — Project a spectrum on a screen as directed in Art. 551. Now place a second prism like the first behind the first, but reversed in position. Observe the character of the image. Slide a piece of cardboard along gradually between the prisms and observe the changes in the image.

Try two prisms reversed in position, one of them crown glass, and the other flint.

555. Exercise.* — Project a spectrum on the screen as directed in Art. 551. Between the prism and screen hold a large convex lens to receive the spectrum. Move the lens along the line of light, and it will be possible to find a position where a white image of the slit is produced on the screen.

If the spectrum is received on a concave mirror a similar result can be obtained.

What does this experiment prove the colored image to be?

556. Exercise.* — Fill an air-thermometer bulb, 4 cm. in diameter, with clear water. In front of the porte lumière, or lantern, place a large white-paper screen in which is a circular opening about 3.75 cm. in diameter. Support the bulb in a clamp so that the cylindrical beam of light is incident on it. There will be formed on the screen a circular spectrum resembling a rainbow. The distinctness of the bow will depend on the distance of the bulb from the screen, and the nearness of the bulb to a true sphere. With sunlight in a very dark room the secondary bow can be seen.

IX. COLOR.

557. Apparatus. — Porte Lumière, Colored Paper, Prisms, Lenses, Cardboard, Color-Top, Colored Glass, etc.

558. Exercise. — Paste a strip of white paper 3 cm. long and 2 mm. wide on a piece of black cardboard several times larger. View this whole strip, placed in a strong light, through a glass prism, holding its edges parallel to the length of the strip. Examine in a similar manner strips of red, blue, yellow, etc., paper. What color or colors do you find in the light reflected by each paper?

559. Exercise.* — Project the solar spectrum as directed in Art. 551. Examine the color of pieces of colored paper when held successively in the different colors of the spectrum. Explain.

560. Exercise.* — Place common salt in the wick of an alcohol lamp and analyze the light emitted by it as in Art. 558. Examine a highly colored picture by its light and account for the appearance.

561. Exercise. — Procure at a toy store a Newton's color-top, or any top having a heavy metallic disk. Cut out of colored paper a disk about 10 cm. in diameter, with a small opening at the centre of the size of the top-handle. Slit them along a radius (Fig. 230), from circumference to centre, so that two or more of them can be placed together, exposing any proportional part of each one as desired. Select seven disks whose colors most nearly represent those of the solar spectrum, put them together so that equal portions of the colors are exposed, and then place this compound disk over the handle of the top when rapidly rotating. Examine the colors in a strong light.

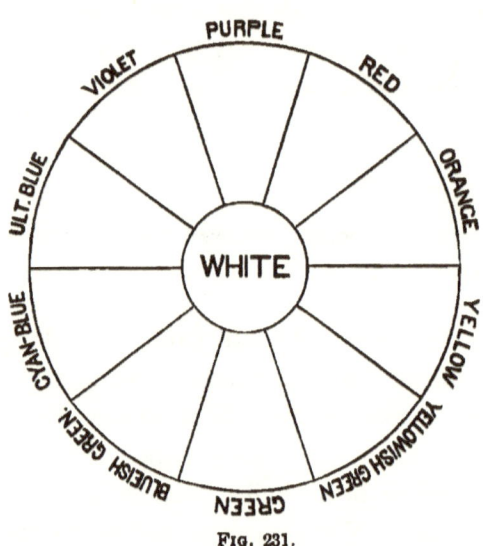

FIG. 230.

FIG. 231.

Try violet, red, and green disks, exposing equal portions. Inference. In like manner try red, green, and blue. Inference.

Try two disks whose colors are opposite in Fig. 231. Inference. Try any two alternate colors. Inference. Ascertain the effect of using unequal portions. Try a black disk and a white one.

A small whirling machine, Fig. 232, or an electric motor, may be substituted for the top.

FIG. 232.

562. Exercise.* — Project on the screen the spectrum of the sun or the electric light. By means of a large lens converge the rays to a focus, producing a white image of the slit on the screen. Prepare two strips of black cardboard, in each of which is cut a slit 2 mm. wide and 2.5 cm. long, and insert them in a strip of blackened wood with a groove cut in it so that the slits can be adjusted at any desired distance apart (Fig. 233). Place this screen between the lens and the prism and set the slits so that one color passes through each opening. Observe the color of the image formed by the lens.

Ascertain if any two colors will produce a white image.

Intercept one of the spectrum colors by a narrow strip of cardboard held by a clamp between the prism and the lens, and note the color of the image.

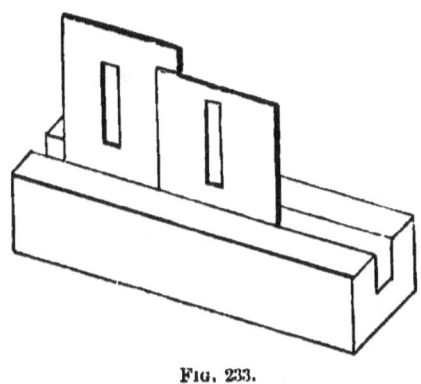

Fig. 233.

563. Exercise.* — Project a bright continuous spectrum on the screen, look steadily at it for a minute, at the end of which time, without removing the gaze from the screen, shut off the light giving the spectrum, and turn up the gaslight or lamp. Observe the change in the image as you continue to look at the screen. Explain.

564. Exercise. — Lay a circular piece of blue paper, 15 mm. in diameter, with a thread attached, on a small sheet of white or gray paper. Look steadily at the blue paper in a strong light for 15 seconds, then, without moving the eye, suddenly pull away the blue disk, and observe the color of the after-image. Try other colors.

565. Exercise.* — Project on a screen with a lantern, or porte lumière, and lens an image of a round hole in a piece of cardboard. Look steadily at the image for 25 seconds, then, without removing the eye, withdraw the cardboard, and observe the color of the after-image.

Fig. 234.

Repeat the experiment with the aperture covered with red glass. Try glass of other colors. Try the chemical tank (Fig. 234) filled with a solution of picric acid. Try other colored solutions.

566. Exercise.*—Paste an image of an object cut out of opaque paper on a piece of colored glass of the size of a lantern slide. Project an image of this object on the screen, gaze steadily at it for a few moments, then suddenly cut off the projecting light, and turn up the gas or lamp. Observe the after-image. Explain.

X. SPECTRUM ANALYSIS.

567. Apparatus.—Porte Lumière, Prisms, Spectroscope, Platinum Wire, Colored Glass, Test-Tubes, etc.

568. Exercise.* — Proceeding as in Art. 551, project the solar spectrum on the screen, using a very narrow slit and a carbon disulphide prism. If the focusing is carefully done, and the room is quite dark, a number of dark lines will be seen crossing the spectrum vertically, showing that the solar spectrum is not continuous. Diagram their position.

These lines are easily seen by placing the eye in the pencil of light diverging from the prism.

569. Exercise. — Turn the slit of the spectroscope (Fig. 218) toward a distant white building. Focus the telescope, and place the prism at the angle of least deviation for that part of the spectrum in the field. It will be found easier to study the spectrum if the telescope tube is thrust through a sheet of cardboard to cut off from the eye the light not passing through the apparatus.

Make a map of a few of the more prominent lines, by taking the reading of the angular scale found on the platform of the instrument when the intersection of the cross-hairs is on the dark line, and laying it off on one side of a rectangle, say 10 cm. long, representing the spectrum. As the ends of the rectangle correspond to the readings given by the angular scale of the spectroscope for the limits of the spectrum, the number of degrees corresponding to each centimetre of the rectangle is easily ascertained, and each dark line located by computing how many centimetres correspond to the angular distance the line is from the end of the spectrum.

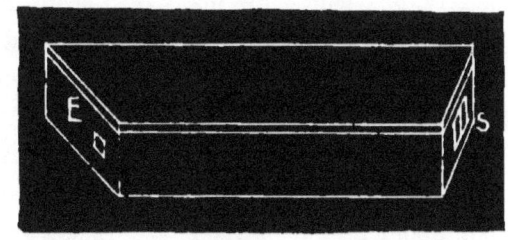

Fig. 235.

As spectroscopes are expensive, the student may construct one by observing the following directions: —

Make out of heavy pasteboard a box 30 cm. long (Fig. 235), whose cross-section is a rectangle. One of the lateral faces should be shorter than the opposite one, as shown in Fig. 236.

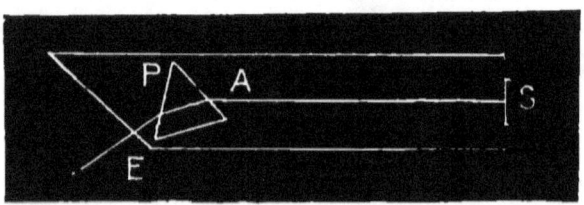

FIG. 236.

In the oblique end cut a small aperture 3 mm. in diameter and considerably nearer the shorter face. In the opposite end cut a rectangular opening 3 cm. square, and cement over it a slit prepared by pasting tin foil smoothly over a piece of clear glass, and cutting a line through the foil 1.5 cm. long by drawing a knife-edge across it. The depth and width of the box will be determined by the size of the prism P. Paint the inside of the box a dead black, using thin shellac varnish and lamp-black. Set the prism in the position shown in Fig. 236, fastening it in place by gluing blocks in the angles. The box must be provided with a tight-fitting cover. The prism described in Art. 551 will show many of the Fraunhofer lines on directing the slit S toward the bright sky and holding the eye close to the opening E, looking in a line perpendicular to that face.

570. Exercise.* — Take a piece of platinum wire of about the diameter of a fine sewing-needle; bend the end into a small loop about 2 mm. in diameter; fuse a small bead of the salt to be experimented with into the loop, and fasten the wire in a clamp so that the bead is brought into the front edge of a gas or alcohol flame at a point a little below the slit of the spectroscope. On looking through the telescope a bright-line spectrum will be seen. Examine in this way the spectra of the following substances, mapping the more prominent lines as directed in Art. 569 : A salt of lithium, potassium, sodium, strontium, barium, etc. Search for Fraunhofer lines occupying the same positions as these bright ones.

571. Exercise.* — Direct the slit of the spectroscope toward the electric light. Place close to the slit a Bunsen flame, and hold in it a pellet of sodium. Mark the position of the dark line crossing the spectrum. Now shut off the electric light, and a bright line will be seen occupying the place of the dark one. Explain.

572. Exercise.* — Observe the spectrum of sunlight when blue glass is placed over the slit of the spectroscope. Try glass of some other color. Ascertain the effect of superposing two or more pieces. Try a test-tube filled with an ammoniacal solution of copper sulphate. Try, successively, solutions of picric acid, quinine, etc. Compare the spectrum with the color of the substance.

573. Exercise. — Place before the slit of the spectroscope a test-tube containing a few clippings of copper. Pour on the copper a few drops of nitric acid, and study the spectrum of sunlight after passing through the reddish-brown gas [1] which will now fill the tube.

574. Exercise.* — Support before the slit of the spectroscope a Geissler's tube charged with some known gas, and observe the spectrum on illuminating the tube with an induction coil.

---

[1] After chemical action has stopped, cork the tube to prevent the fumes from escaping and corroding the metallic parts of the spectroscope.

XI. INTERFERENCE OF LIGHT.

575. Apparatus. — Porte Lumière, Thick Glass, Iron Clamp, Window-Glass, India-Ink, Lenses, Nobert Grating, Perforated Cardboard, Mother-of-Pearl, etc.

576. Exercise.* — Procure two square pieces of heavy plate glass, about 8 cm. on an edge. Clean them thoroughly, then press them gently together by means of common spring clothes-pins on three of the corners, and the supporting clamp at the fourth corner. Let a beam of sunlight be incident on the surface at an angle of 45°. In the reflected beam place a lens, and project an image of the porte-lumière opening on the screen. Notice the colors on the image as the pressure at the clamp is changed. Explain.

Beautiful colors are seen if two such pieces of glass are firmly pressed together at their centres by means of a small iron clamp.

577. Exercise.* — Make out of coarse iron wire a ring 8 cm. in diameter, soldering it to a wire for a handle. Dip the ring into a soap solution prepared as directed in Art. 80, and support it in the beam of light in the place of the glass plates in the last experiment. Observe the play of colors across the image on the screen.

Observe the phenomenon when one-colored light is employed.

578. Exercise. — Paint one side of a strip of window-glass with India-ink, rendering it opaque. When dry, rule, with a fine needle, a number of parallel lines, as close together as possible, by cutting through the ink to the glass.

Now stand about 5 m. away from a lighted lamp and view it through this grating. Explain.

View through this grating a slit placed across the porte lumière.

579. Exercise.* — Project with a lens an image of a slit placed across the porte lumière, having the room as dark as possible. Over the slit place a Nobert grating with its lines parallel to the slit. Observe the series of spectra. Compare the relative spaces occupied by the colors with the relative spaces when the spectrum is obtained by a prism.

Photographs of these gratings are nearly as efficient as the originals, and cost only about one-fourth as much.

580. Exercise.* — Reduce the opening in the porte lumière to a circular hole 2.5 cm. in diameter. Cover this aperture with a piece of perforated cardboard, and project with a lens an image of it on the screen. Try two pieces, slowly sliding one over the other. Explain.

581. Exercise.* — Substitute for the glass plates in Art. 576 a piece of mother-of-pearl, and place a slit in the opening of the porte lumière. A peculiarly colored image should be obtained. Now substitute for the pearl an impression of it taken in black sealing-wax. Explain.

XII. OPTICAL INSTRUMENTS.

582. Apparatus. — Florence Flask, Lenses, Cardboard Telescope, etc.

583. Exercise.* — Fill a $\frac{1}{4}$-litre globular flask with distilled water to represent the eyeball. Cover one-half of the globe with black paper, with a round hole cut in it. This will represent the iris and pupil. Place a convex lens of long focal distance in front of this hole to represent the cornea and the crystalline lens combined. Support a lighted candle in front of this hole, and at such a distance as to form a clear image on a paper screen placed behind the globe. Now move the candle toward the globe till the image is quite indistinct, and then endeavor to restore the distinctness of the image by interposing a second convex lens. Also move the candle away from the globe, and restore the distinctness of the image by interposing a concave lens. What defects of the eye are illustrated in these experiments?

584. Exercise.* — The principle of the microscope may be illustrated as follows: Select two convex lenses of the focal distances 3 and 5 cm. respectively. Mount them carefully in cork or wooden rings, as shown in Fig. 237, cutting a small portion from one edge to produce a flat surface. Glue these mounted lenses on a small board 15 cm. long by 3 cm. wide, placing the one of longer focus at the end A, to serve as the

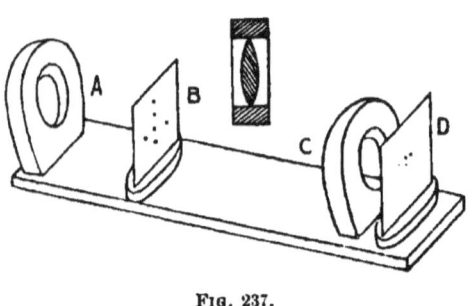

Fig. 237.

eye-glass, and the other distant from it 11 cm., as at C, to serve as object-glass. Cut two cork disks, B and D; insert into one of them a rectangular piece of stiff writing-paper, and into the other a similar piece of thin sheet-brass. Slightly oil the paper to render it translucent, and drill in the brass plate a number of small holes, arranged in the form of a cross, at the height of the centre of the lens above the base. Before placing the brass plate D in position, place the paper screen B about 4 cm. from the lens A, and move it to and fro till a position of distinct vision is found for it when viewed through A. Now place a flame at the other end of the board, exactly in line with the centre of the lenses, and between it and C place the brass plate D, distant from C about 17.5 mm. Move D till a distinct image of the cross is formed on B; then a magnified image of the cross will be seen through A, but more clearly defined if B is removed. What is the office of each lens?

To determine the magnifying-power ascertain, as in Art. 549, the magnifying-power of the eye-piece and object-glass separately, and obtain their product.

585. Exercise. — The principle of the telescope may be illustrated as follows: Mount two convex lenses as directed in the last article, having the focal distances 30 cm. and 5 cm. respectively. Cement the larger one, A, to one end of a board 50 cm. long by 10 cm. wide (Fig. 238), and the other, B, to a second board 5 cm. square, but of such a thickness as to bring the centres of the lenses in a line parallel to the large board when the small board

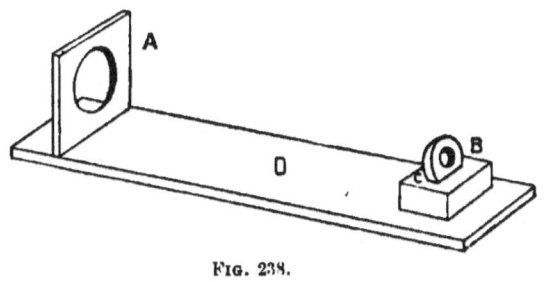

Fig. 238.

rests upon it. Prepare a translucent screen as in the last article. Place the apparatus on one end of a long table, with the lens A directed toward a candle-flame at the other end. Set the screen to receive the image of the candle formed by A;

then by moving the eye-piece B, a place will be found, giving an enlarged image of this image on the translucent screen, and a brighter image of the candle on removing the screen. What is the office of each lens?

If the eye-piece is made of two lenses, b and c (Fig. 239), b having a focal distance of 5 cm. and c of 3 cm., the distance between them being 9 cm.; on placing the screen between b and c, an erect image of the candle will be obtained instead of an inverted one as before, and on looking through b an erect and magnified image of the candle will be seen. What is the office of the lens c?

Fig. 239.

586. Exercise.* — The principle of the Galilean telescope may be illustrated as follows: Substitute for the lens B in the last experiment a concave lens of 5 cm. focal length. Distinct vision will be secured on placing B between the screen and the lens A.

587. Exercise. — Determine the magnifying-power of a telescope.

Direct it toward the open sky, and there will be seen near the eye-piece, and a little beyond it, a small illuminated circle. This is an image of the objective opening of the telescope. Measure the diameter of this image, employing a finely divided scale. The magnifying-power is equal to the quotient of the diameter of the object-glass by that of this illuminated circle.

XIII. DOUBLE REFRACTION AND POLARIZATION OF LIGHT.

588. Apparatus. — Iceland Spar Crystal, Tourmaline Tongs, Polariscope, Lenses, etc.

589. Exercise. — Place a crystal of Iceland spar over a small dot on a white sheet of paper, and observe the dot through it. Rotate the crystal about an axis perpendicular to the paper, and notice the behavior of the images of the dot. Mark the relative brightness of these images while the crystal rotates as you view them through a tourmaline plate or Nicol prism.

590. Exercise. — Procure a pair of tourmaline tongs, an instrument consisting of two thin plates cut from the tourmaline crystal and mounted as shown in Fig. 240. Look through the plates at the bright sky and observe the changes in brightness as one plate is turned on an axis within the supporting ring.

Fig. 240.

Set the plates so that the view is darkest, then place between them a quartz spectacle-lens, known as a pebble-lens, comparing the effect with that produced by a glass lens.

Opticians prepare thin plates of crystalline substances, such as Iceland spar, quartz, arragonite, etc. Place any of these you may be able to procure between the plates of the tongs and observe the peculiar figures and the play of colors as one of the plates is rotated.

591. Exercise. — Examine with a polariscope pieces of unannealed glass, as glass stoppers, paper weights, Rupert's drops, etc.; also thin plates of mica, selenite, quartz, Iceland spar, etc.

An efficient polariscope may be constructed as follows: Cut a rectangular piece of board 36 cm. long, 10 cm. wide, and 2.5 cm. thick, and also a right triangular one, having the sides about the right angle 24.5 cm. and 17.4 cm. respectively.

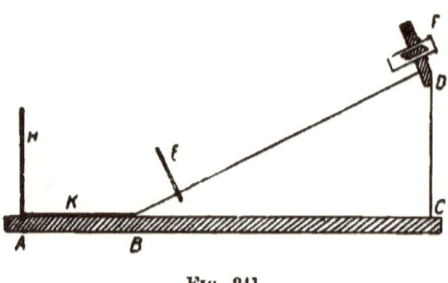

Fig. 241.

Fasten these pieces together as shown in Fig. 241, the shorter side about the right angle being the vertical one. At A, 5 mm. from the end of the board, cut a groove to support a plate of ground glass, H, 10 cm. square. Between H and B place three or four pieces of thin plate glass, each 10 cm. square, on a piece of dead-black paper to serve as a polarizer. At E, 4 cm. from B, support in a transverse groove perpendicular to the plane of BD, a glass plate, to serve as a shelf for supporting the objects under examination. At D glue a wooden block 5 cm. high, and 2.5 cm. thick, having an aperture through its centre 2.5 cm. in diameter, parallel to the surface of the plane, to support the analyzer. This analyzer may be constructed in either of two ways: first, by procuring from an optician a small Nicol prism, and mounting it in a centrally apertured cork inserted in a paper or brass tube, 2.5 cm. in diameter (Fig. 242, A); or, secondly, by procuring a paper or metal tube as before, fitting to it an inner tube of pasteboard divided obliquely (Fig. 242, C), at an angle of 35° 25′ with the axis of the tube, and placing between these two parts 12 or 15 elliptical microscope cover-glasses, shown in section in Fig. 242, B. In either case one end of the tube is capped, and in the cap is a central opening 2 mm. in diameter. A shoulder is provided for the tube by gluing on the outside a paper ring.

In using the instrument let light from the bright sky pass through H, and be incident on K, by which it is reflected through the object on E, to the analyzer F. On rotating F on its axis, a fine play of colors will be seen on the object, due to interference.

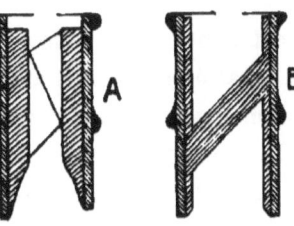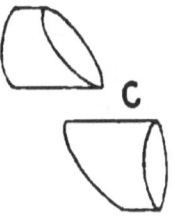

FIG. 242.

A very convenient instrument for investigations in polarization is Norremberg's doubler, a simple form of which may be constructed as follows: Knock out the opposite sides, EL and PM, of a large cigar-box, and on the other two lateral faces fix guides for the polarizing plate B (Fig. 243), so that the plate makes an angle of 35° 25' with the edge EF. On the end FM lay a piece of good looking-glass the size of the end; and in the other end cut a hole, in which the analyzer A, either a Nicol prism or a bunch of glass plates, can be rotated. The objects under examination are placed on the looking-glass. The light incident on B is reflected to D and partially polarized; D in turn reflects it through A to the eye. If desired, a horizontal shelf between B and A could be put in, having a circular opening for the passage of light. Objects could then be examined resting on this shelf over the opening.

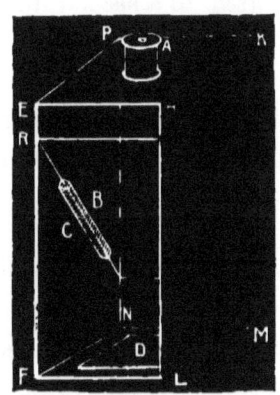

FIG. 243.

592. Exercise. — Procure several strips of glass, such as are used in preparing microscope slides. Form on the centre of each a paraffine ring 2 cm. in diameter. Support the glass in a horizontal position, and put within the circle a few drops of a solution of some chemical salt, and let it remain undisturbed until crystals are formed. Examine these under a microscope with a polarizing attachment. The following list of salts is recommended: Alum, potassium bichromate, mercury bichloride, boracic acid, potassium carbonate, citric acid, potassium chlorate, potassium iodide, ammonium nitrate, copper sulphate, iron sulphate, potassium sulphate, potassium ferrocyanide, etc.

An instrument suitable for examining such crystals can be constructed as follows: Construct an analyzer as directed in the last article, and fit it in the draw-tube of a compound microscope (Fig. 244), between the eye-piece and the object-glass. A similarly constructed tube must be made for a polarizer, and fitted to a second tube somewhat shorter, turning freely within it without falling out. Cement this outer tube to the under side of the microscope stage with the axis of the tube exactly in that of the draw-tube above, but leaving the inner tube free to turn on its axis. Very satisfactory results can be obtained from the common draw-tube microscopes, and even the simple botanizing glass can be easily adapted to exhibit the phenomena of polarized light (see "The Scientific American" for July 31, 1886). Paper tubes to fit the tubes of the microscope are easily made by gumming. writing-paper, and winding it around a cylinder of the proper size.

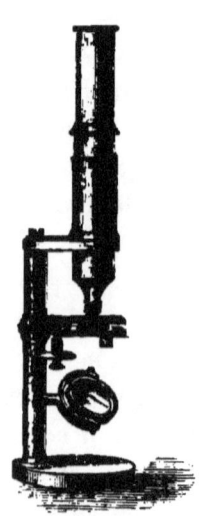

FIG. 244.

593. Exercise. — Dissolve salicine in one part alcohol to four parts water, made rather hot, obtaining a saturated solution. Pour a layer on a microscope slide, and evaporate quickly with rather a strong heat. Examine the crystals obtained under the microscope with the polariscope attachment.

APPENDICES.

APPENDICES.

APPENDIX A.

THE PHYSICAL LABORATORY.

594. The Room. — Where a separate room can be provided for experimental work in Physics, it should be large, well ventilated, and have a southern exposure. Opening off from it should be a room for the storage of apparatus. Provision should be made for darkening the windows, either by inside blinds or opaque curtains mounted on spring rollers. If curtains are used, the lateral edges must slide in deep grooves, made by nailing strips on the casing, to prevent the light from entering the room around the edges.

A few heavy *flat-top tables*, about 32 in. high, 3 ft. wide, and of such length as the room permits, should be obtained. These tables should have no iron in their structure, and to make it possible to clamp apparatus on them the top should project at least four inches beyond the frame. In a part of the room where there is a good light, place a carpenter's *workbench*, with a *vise* and *anvil* upon it. Above it, on the wall, fasten a cupboard, to contain such tools as *saws, planes, brace* and *bits, drills, hammer, chisels, try-square, files, wire-cutter, soldering-iron, oil-stone, nails, screws*, both *iron* and *brass, tacks*, etc. This part of the equipment should not be overlooked. School authorities will seldom refuse to appropriate money to be spent in this way.

A side-table for the few chemicals needed will be found convenient. A *sink* for waste water must by no means be omitted.

Connect the room with the water system if there is one; if not, place a galvanized iron tank at one end of the sink, having a capacity of two or three barrels, to be filled with water from the well or cistern by means of a force-pump.

As soon as possible put a *well-filled bookcase* in the room, with a table and chairs near by, where students may look up points regarding their work. Such books as the following are recommended: —

Ganot's Physics, by Atkinson; Practical Physics, by Guthrie; Cooke's Chemical Physics; Elementary Practical Physics, by Stewart and Gee; Practical Physics, by Glazebrook and Shaw; Physical Manipulations, by Pickering; Experimental Physics, by Weinhold; Physical Constants, by Everett; The Art of Projecting, by Dolbear; Light, by Wright; Light, by Mayer and Barnard; Sound, by Mayer; Sound, by Tyndall; Magnetism and Electricity, by Guthrie; Electrical Rules and Tables, by Munro and Jamieson; Fleming's Short Lectures to Electrical Artisans; Practical Mechanics, by Perry; Physical Measurements, by Kohlrausch; Practical Electricity, by Ayrton; Handbook of Electrical Testing, by Kempe.

The walls of the room would look more attractive if they were decorated with the pictures of prominent scientific men. Bare walls are not very inspiring; good work is not wholly independent of the student's surroundings. Charts of spectra neatly framed would be both useful and ornamental.

If illuminating gas is procurable, by all means introduce it into the laboratory, supplying each table with jets suspended from the ceiling, making connections with them by rubber tubing. The advantage of this method is that the tables can be moved about if necessary; there is no iron beneath them to interfere with the action of magnetic needles, and when the gas is not in use the fixtures are out of the way.

It is not to be expected that every luxury, or even every convenience, can be provided at the outset. Let the essentials be secured, and then add from time to time such fixtures as will enable the work to be done with greater accuracy and at less inconvenience.

If the recitation-room is large, and the class small, the back part of the room might be fitted up for experimental work. This would be much better than the entire omission of this important kind of training. The days of teaching science at long range, as it were, are passed. That the pupil must come in actual contact with the phenomena to be studied is recognized by every wide-awake instructor. The apparatus should be put in his hands, with clear directions regarding its use; in this way there are trained all of the powers of the mind, the observational not the least important.[1] Faraday used to affirm that he always wished to perform experiments himself; that he always learned something in doing the work that the description in words could not convey to him. An experiment which fails is often more instructive than one which succeeds. In finding the cause of failure more will often be learned than when the work goes through, as if by routine.

595. Apparatus. — It is a mistake to suppose that elaborate and highly finished apparatus is necessary for successfully prosecuting experimental work in Physics. Neither should it be expected that all the required appliances for the complete investigation of every department of the subject can be pro-

---

[1] A few experiments performed by himself will give the student a more intelligent interest in the subject, and will give him a more lively faith in the exactness and uniformity of nature, and in the inaccuracy and uncertainty of our observations, than any reading of books, or even witnessing elaborate experiments performed by professed men of science. J. CLERK MAXWELL.

vided at the outset. Let some of the more important pieces be secured, selecting them with the view of adding the others as soon as practicable. In this way an outfit will be obtained in time, where each piece has its place and all fit harmoniously together. Much can be done with a few wisely chosen pieces. Although the author does not believe that giving instruction in physical technics is the proper work of the laboratory, yet he would permit a limited amount of apparatus construction, if too much time is not consumed in it. Each class will usually have its handy boy, whose mechanical gifts can be turned to good account by having him construct, now and then, a desirable piece of apparatus. The school janitor is generally a man of some mechanical skill, and will willingly construct frames, supports, and the like. A small fee levied each term, to cover incidentals, will always be cheerfully paid by the patrons of the school, and in the course of the year will amount to a handsome sum, to be devoted to improving the equipment.

Among the first things to be purchased are appliances for measuring Metre sticks can be procured of the Metric Bureau for 25 cents each; several of these should be obtained. For measuring small linear quantities, where a selection from the instruments described in the first chapter has to be made, it is recommended that the *Dividers*, *Diagonal Scale*, and *Micrometer Caliper* be chosen, and that several of each of the first two be purchased. The *Circular Protractor* will be needed for angular measurements. Cheap and efficient ones are now made of horn.

Good reliable *Balances* are indispensable to every laboratory, but unfortunately they are expensive. The *Jolly Balance* can be constructed by any one at all skilful with tools, and makes a very good substitute for the beam balance, provided accurate weights are used. If a support is fitted to the *Jeweller's Hand*

Balance, a good deal of work can be accomplished with it, if one is very careful. Balances sensitive to a centigramme can be imported, duty free for schools, from Germany for about $6.50. With *one set* of *Accurate Weights* for comparison, cheap ones can be adjusted, and, in fact, it is not a difficult matter to make weights out of sheet-brass, brass wire, and aluminum wire.

Of *Glass Tubing* there should be a *liberal supply* of both large and small sizes. Select tubes having thick walls, as they are stronger, and bend without collapsing. *Rubber Tubing* for connections, and for many other purposes described in the preceding pages, is one of the necessities.

A few shillings appropriated for wire, brass, copper, and iron, of various diameters, will well repay. For electrical work covered copper wire will be required; the desirable numbers are indicated in the chapter on electricity.

In constructing supports and many simple devices, there will be frequent use for *well-seasoned pine lumber* of different thicknesses. Material of this kind should be always on hand.

If gas is available, the best lamp for heating purposes is the *Bunsen Burner* (Fig. 245), as it gives a very hot and smokeless flame. In using the burner care should be taken to regulate properly the supply of air through the holes near the base, for if too much air is admitted, the flame is likely to strike down the tube and burn at the lower part, giving a smoky flame, the heat of which is expended, in a measure, on the tube. When gas is not available, resort must be had to the alcohol lamp, or to some form of gasoline burner. An alcohol lamp can be extemporized out of a bottle, to which is fitted a perforated cork, through which passes some candle-wick. The top of the cork can be protected from the flame by a layer

FIG. 245.

of plaster-of-Paris. In Fig. 246 is shown the *Mouth Blowpipe*, an instrument easily made from two pieces of glass tubing and a sound cork. When great heat is required, as in manipulating heavy glass tubing, resort must be had to the *Blast Lamp* (Fig. 247). Fig. 248 shows a very convenient form of *Blower* for supplying air to the blast lamp.

Fig. 249 represents the *Florence Flask*. The laboratory should be supplied with several of these, varying in size from a half-litre to a litre and a half. When heat is to be applied to one, place it in a sheet-iron shallow saucer, containing fine sand, or a square of asbestos paper, supported on a ring of the iron stand (Fig. 262), to secure a uniform distribution of heat. The *Glass Retort* (Fig. 250) may often take the place of the flask, especially when any distillation is to be done. Fig. 251 exhibits a very useful form of *Condenser*. A large bottle or flask may usually be used as a substitute. *Funnels* (Fig. 252) are almost indispensable. They may be of either glass or tin. One form is provided with a stop-cock. The *Graduate* (Fig. 253) is employed for the volume measurement of liquids. These

Fig. 247.

Fig. 248.

Fig. 249.

Fig. 246.

THE PHYSICAL LABORATORY. 343

instruments should be graduated in both the English and the metric system. At least two sizes will be desirable, one having a capacity of 25 ccm. and graduated to centimetres, and a second one holding at least 500 ccm. The *Cylindrical Graduate* (Fig. 254) cannot well be dispensed with, being almost a necessity in determining the volume of irregular solids by the displacement of water. These should be graduated to cubic centimetres. Two or three of these should find a place in the laboratory, and may be used in place of the conical form. Fig. 255 shows the *Pipette-Tube*, and the manner of using it. A glass tube, drawn out a little at one end to narrow the opening, makes an excellent substitute. *Beakers* (Fig. 256) are to be had of various sizes and shapes. A few holding a litre, and several of less volume should be provided. Copper beakers are very desirable, as there is no danger of breaking them by sudden changes of temperature. The first outlay is somewhat greater, but it pays in the end to procure them. A dozen, at least, of large *Test-Tubes* will be needed, and two or three times as many

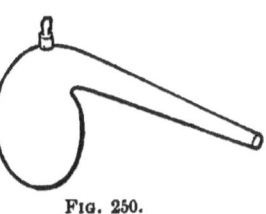

FIG. 250.

FIG. 251.

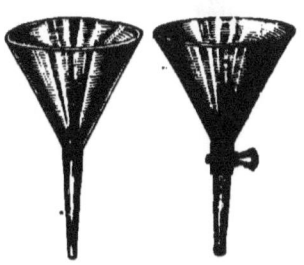

FIG. 252.

FIG. 253.

of smaller size (Fig. 257). To clean them, a specially constructed brush (Fig. 258) is recommended.

It is frequently desirable to reduce a substance to a fine powder. In that case a *Mortar and Pestle* (Fig. 259) is necessary. If the substance to be powdered is not very hard, a stout bowl might be used with a pestle made out of hardwood. Very hard substances must be broken with a hammer or in an iron mortar. For condensing solutions or drying substances use an *Evaporating Dish* (Fig. 260). They are made of either glass or porcelain. Common white stoneware saucers are an excellent substitute, and are quite inexpensive. In heating them, the temperature must be increased gradually, and they should never be exposed to the naked flame, but should be set in a dish of sand and kept at a moderate heat. Where there is danger of overheating during evaporation, or where it is necessary to keep a substance at a constant moderate temperature, the *Water-Bath* (Fig. 261) is recommended. This

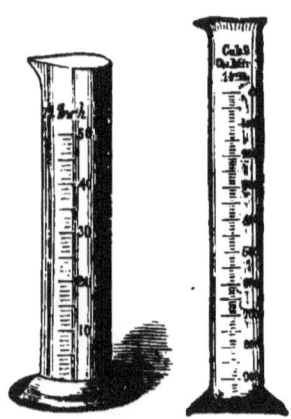

Fig. 254.

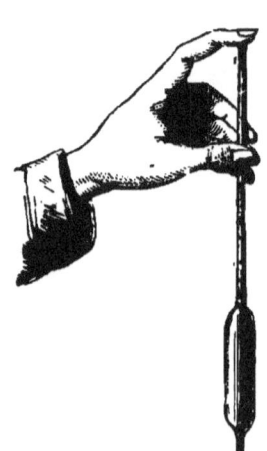

Fig. 255.

Fig. 256.

THE PHYSICAL LABORATORY. 345

is only a copper basin provided with several rings of different sizes, that the top may be adjusted to the dish to be placed upon it.

Figs. 262 and 263 illustrate the *Iron Stand* and the *Universal Holder*, respectively. Each student, or set of students, working together, ought to be supplied with one of each, and

FIG. 258.

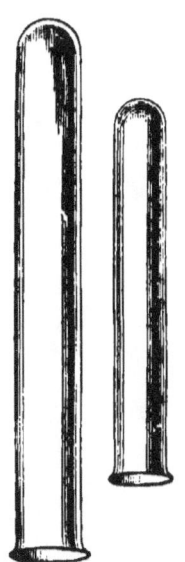

occasionally more than one of the latter will be found convenient. The price of the iron stand varies with the number of rings supplied with it. For most work the single ring will be sufficient. The universal holder can be made by any good mechanic, and in this way secured at a price within reach. The Burette holder differs from

FIG. 257.

FIG. 259.

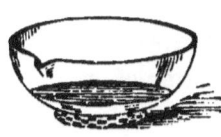

FIG. 260.

FIG. 261.

the universal in being adjustable only to height. It is much less expensive, and answers equally well in supporting lenses, cardboard screens, jet-tubes, etc. Fig. 264 exhibits a good form of *Test-Tube Rack*, and Fig. 265 a *Test-Tube Holder*, the latter to be used when heating a solution in a flame. In handling small heated vessels, or heated substances, the *Crucible Tongs* (Fig. 266) will be found convenient. The adjustable *Table Support* (Fig. 267) will be found useful in supporting beakers, prisms, etc., but is not indispensable. Fig. 268 represents the *Tinner's Nippers*, a useful instrument for cutting metals in sheet form. Fig. 269 is a *Clamp* for closing rubber tubes. Fig. 270 shows a *Cork Borer* and the manner of using it. A small round file makes a good substitute. In reducing the size of a cork, use a flat coarse-cut file. To prevent small corks splitting when drilling holes through them, wrap them firmly with several turns of stout twine. In perforating rubber corks, keep the borer wet with a solution of caustic potash.

In electrical work a few good battery cells are needed. These should be selected in view of the kind of service required of them. For intensity, where constancy is immaterial, there is nothing better than the

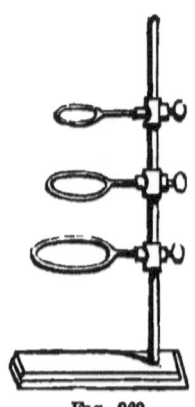
Fig. 262.

Fig. 263.

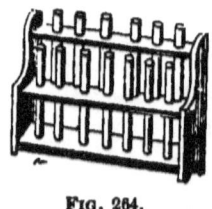
Fig. 264.

Fig. 265.

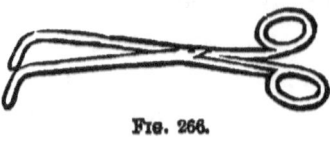
Fig. 266.

Fig. 267.

Grenet Battery (Fig. 271), charged with the chromic acid solution. For constancy, where strength of current is not so much of an object, the *Daniell's Battery* (Fig. 272) is superior to most others. The forms of batteries are legion, but for cheapness and efficiency there is none superior to the two mentioned. It is too expensive to rely upon batteries for such currents as are required for electric lighting. Here resort must be had to the *Dynamo*. Small ones are now to be had giving one small arc light, suitable for optical

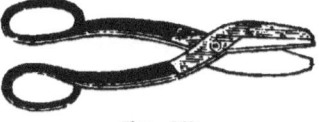

FIG. 268.

FIG. 269.

FIG. 270.

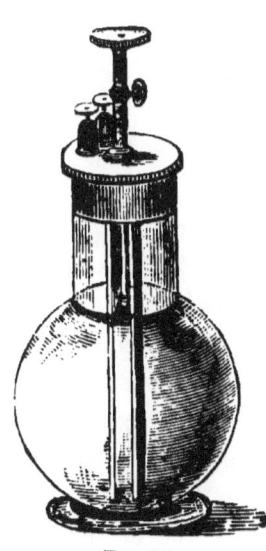

FIG. 271.

FIG. 272.

work, for about $100. Mr. Geo. M. Hopkins, in the Supplement to the "Scientific American," No. 600, gives full directions for constructing such a dynamo. Where water or steam power is available, there is no more important instrument for the physical laboratory than the dynamo, on account of the strong light that can be obtained from it. Sunlight is so uncertain, that if optical work is made dependent on such a source of light many a disappointment will be met with, and considerable disorganization of the work will result therefrom.

596. Conducting the Work. — Probably the greatest drawback to carrying on experimental work in physics is the apparent impossibility of accommodating such large numbers of students as are always found desirous of studying the subject. When the class is very large it is recommended that it be divided into five divisions, one division to meet for work each day for, say, two hours, working in sections of two pupils each. The problems should be assigned a few days in advance, that they may be carefully studied in all their parts by the pupils before entering the work-room, so that no time will be lost in ascertaining what is to be done, and in what manner. It is also recommended that they be reviewed by the instructor, before the class, before the pupils are set to work upon them. Each pupil should have his appointed place in the room, where his work should all be done, except where the nature of it requires a change. The apparatus needed for his work should be placed on his table in advance, say, the Saturday before, remaining there during the week for the use of the different sections. When duplicate pieces of apparatus are not on hand, and several pupils require the same piece at some stage of their work, a special table had better be provided for such appliances where they may be found when needed. This is more especially true of the air-pump, whirling-machine, Atwood's machine, electrical machine, etc. On completing an experiment the pupil should be required to make carefully worded notes on all that he has observed. This record should be submitted for approval immediately after completing the experiment, and if found correct, as well as complete, in all its parts, then the next experiment can be entered upon; but if unsatisfactory, the pupil should return to the same experiment, and repeat it till all errors are corrected and omissions supplied as far as possible.

When the classes are small, or assistance is furnished the instructor, the number of hours devoted to experimental work should be increased. Four or six hours per week spent in the laboratory, and perhaps three in the recitation-room, drilling on principles and reviewing the facts developed by experiment, when carried on for a school year, will demonstrate that there is no study in the curriculum that surpasses physics in the breadth of mental discipline imparted.

It is not expected that each pupil will attempt all the work described in this book, but a judicious selection should be made, giving each pupil some work under each topic, the amount to be determined by the time available, the aptness of the pupil, and the laboratory facilities. Many topics might be omitted entirely, and if found desirable the order of the others might be changed, if by so doing there is secured a closer conformity to the order found in the text-book used in the class-room.

The following list of articles selected from the preceding pages is suggested, as offering a fairly complete and easy course, supplementing to good purpose the experiments of the recitation-room : —

2, 4, 5, 9, 12, 13, 18, 20, 24, 25, 31, 62, 64, 67 to 70, 73, 78, 79, 83, 84, 108 to 123, 137 to 148, 151 to 163, 182 to 197, 216, 219, 222, 234, 239, 241, 246, 247, 253, 254, 269 to 274,

285 to 300, 306 to 312, 315 to 322, 325, 326, 331, 346 to 353, 358 to 361, 378 to 382, 386, 387, 391, 412 to 416, 428, 429, 431, 434, 441 to 443, 452, 454, 456, 472, 476, 483, 485, 491, 492, 504, 511, 512, 516, 517, 521, 522, 528, 530, 532, 534, 539, 541, 545, 547, 548, 558, 561, 564.

In response to numerous inquiries from teachers, concerning physical apparatus, and where it can be bought at fair prices, the author would take this occasion to say that anything described in the preceding pages can be furnished by either Eberbach and Son, Ann Arbor, Mich.; E. S. Ritchie and Sons, Brookline, Mass.; A. P. Gage, Boston, Mass.; or Jas. W. Queen & Co., Philadelphia, Pa. By writing to either of these houses, and referring to the articles and figures of this book, you can promptly learn at what prices the various appliances can be obtained. The author will gladly give any information in his power regarding apparatus, answer any questions as to its use, and aid teachers in any way he can in overcoming difficulties encountered in endeavoring to teach physics on the plan outlined in these pages.

597. The Note-Book. — Every pupil must supply himself with a note-book in which to record the transactions of the laboratory. This book should be a model of neatness; work slovenly done should be rejected, and the pupil should be required to rewrite it. The use of loose sheets of paper for note-taking should not be permitted. Calculations and sketches should be shown on the left-hand page, and observations, descriptions, formulæ, and theory on the right-hand page. During the evening following the day devoted to experimentation, the pupil should prepare a full report of his day's work, describing the character of each experiment, the nature of the apparatus employed, the results obtained, and the inferences which he thinks can be deduced from them. After these reports have been examined by the instructor, and the errors which were pointed out corrected by the pupil, they should be neatly copied into a second book provided for the purpose. One recitation-hour per week may be very profitably devoted to the consideration of these reports in the presence of the whole class, the instructor passing upon their accuracy, completeness, and literary execution.

598. The Graphic Method. — This method is frequently employed to find out the probable form of the law connecting two quantities which are so related that a change in one is attended by a change in the other; as, for instance, in Art. 74, where an increase in the weight placed in the pan produces a change in the length of the wire, and in Art. 88, where a change in the temperature of the solvent is attended by a change in the solubility of the substance. It consists in representing by a line the data obtained by the observer in the course of his investigations, the shape of the line indicating the form of the law connecting the dependent facts. To construct the line, procure a piece of cross-section paper of suitable size. This is merely a good quality of writing-paper divided into equal small squares, by a great number of horizontal and vertical lines, the most desirable sizes for these squares being 1 mm. or .1 inch on a side. Such paper can be purchased of any dealer in mathematical instruments. Now select some suitable scale to be used in representing the things to be compared, the chief condition being that the units be such as to bring all the work on one sheet. For instance, in Art. 74, the number of units of weight used each time may be represented by as many divisions laid off on one of the horizontal lines, the side of a square serving as the linear unit. If this is found to require too long a line, the side of a square may represent two or more units of weight; and, on the other hand, if the line is too short, a larger unit may be chosen. In like manner any convenient unit may be selected to represent the elongation of the wire. After suitable units have been decided on, not necessarily the same for both quantities, then measure toward the right on the horizontal line, passing through the corner of some one of the squares (usually taken near the lower left-hand corner of the sheet, and called the *origin*) a sufficient number of spaces and parts of a space to represent the numerical value of the first of the dependent quantities, and then, in the perpendicular through the point just located, measure off a distance to represent the numerical value of the second quantity, marking with a cross the point thus located. In like manner, locate points, using the remaining data, making all measurements from the same origin, and employing the same units. Negative data should be measured off in directions exactly opposite to those used for positive quantities. Through the points located, sketch as smooth and symmetrical a curve as possible. The greater the number of points, and the closer they are together, the better it will be for sketching the curve. The simpler the line passing through these points, the less complex the law connecting the quantities; as, for instance, a straight line would imply that one quantity is proportional to the other. On account of the errors which unavoidably enter into all observations, the points will not all lie exactly on a smooth curve, but may be a little to one side of it. Thus, the Graphic Method often portrays to the eye just where mistakes have been made by the observer, and also enables him to get approximate values for intermediate points in the problem. See Stewart and Gee's Practical Physics, Vol. I., page 275.

599. Summary of Laboratory Rules. — Pupils are advised to observe carefully the following : —

1. Be orderly and neat in manipulation. Keep all apparatus clean, never setting away any apparatus without wiping it with a chamois, if metal, and if glass, washing it and thoroughly drying it.

2. Economize time by remembering that frequently two or more operations can be carried on simultaneously; never, however, if either one needs close and constant attention during any considerable period of time.

3. Prepare thoroughly for every experiment, so that every condition will be observed, and no failures result from a neglect of some requisite. Do not substitute for what is the best possible, that which will barely do.

4. Ascertain the reason for every step taken in working through an experiment, noting which are essential conditions permitting no variations, and which are non-essential only in so far as they conduce to greater accuracy, and hence may be modified as lack of apparatus may compel.

5. Be exact and methodical. Let nothing pass unnoticed, although you may not see its significance at the moment.

6. Keep your note-book by you, and record in it carefully, neatly, and at the time the results of the observations, and also make sketches of the apparatus used.

7. *Do not crowd the notes;* leave plenty of room after each experiment to write out a full explanation of the facts, together with the principles revealed by them. If several similar quantities are recorded, arrange them in tabular form with parallel columns.

APPENDIX B.

LABORATORY OPERATIONS.

600. Cutting Glass. — To cut small glass tubes, make a deep scratch with a three-cornered file across the tube as it rests on the table; then holding the tube, as shown in Fig. 273, with

Fig. 273.

the scratch turned from you, press suddenly outward with the thumbs, and it will break off as desired. To cut off large tubes, flasks, bottles, etc., make a deep scratch with a file, then apply to it either a pointed piece of glowing charcoal, a heated metal rod, or a fine gas-jet. The sudden expansion by heat will generally produce a crack; if not, then touch the heated spot with a wet stick. A crack thus started may be led in any desired direction by keeping the source of heat moving slowly a few millimetres in front of it as it advances. To get a small-pointed gas-flame for the above purpose, connect a glass jet-tube to the gas-supply by a piece of rubber tubing. Another way to cut a glass cylinder is to wind a turn of fine platinum wire around it, just where the scratch has been made, bringing the ends as close together as possible without touching. Now pass through the wire an electric current of sufficient intensity to bring the wire to a white heat. Glass plates are readily cut with the common "glass-cutter," an instrument provided with a highly-hardened steel wheel in place of a point.

601. Smoothing the End of a Glass Tube. — Warm the end by passing it to and fro through the Bunsen, or spirit flame; then hold it obliquely in the flame, with the end just inserted, slowly turning the tube around. Remove from the flame soon after the flame becomes a bright yellow. In the case of large tubes it will be necessary to smooth them by grinding on a smooth flag-stone wet with water.

602. Bending Tubes. — Warm the tube for several inches each side of the place where the bend is to be by slowly passing it through the flame; then bring the tube into the gas-flame, slowly rotating it on its axis, heating it evenly for about three centimetres. When the flame becomes a bright yellow the tube will be soft. Then

FIG. 274.

gently bend the ends from you until the required angle is obtained. Fig. 274 shows how to hold the tube. In bending moderately large tubes a flat flame is preferable. Fig. 275 exhibits an attachment for the Bunsen burner to produce such a flame. The fish-tail gas-burner or the

FIG. 275.

kerosene lamp gives a very suitable flame for use in bending small tubes.

603. Closing the End of a Tube. — Soften the tube at the end by holding it in the flame, and then pull the end out with another piece of glass. Keep removing the small tail that is left till it becomes quite small; then heat the end for a few minutes, turning the tube in the fingers. If the tube is held in an inclined position in the flame, the opening will keep contracting till finally it closes up.

604. Drawing out Tubes. — Thoroughly soften the tube uniformly for three centimetres of its length, then remove from the flame and pull the parts asunder till the diameter is about 1 mm., holding the tube steadily till it cools, to avoid crooking it. Now scratch it with a file at the smallest part and break it in two, smoothing the ends in the flame.

605. Drilling Holes in Glass. — Small holes are quite readily drilled through glass by using a hard drill, wet with a solution of camphor in oil of turpentine. A three-cornered file, with the end broken off, makes a good drill for glass. Large holes may be drilled with a brass tube and emery powder moistened with water. A piece of wood, with a hole in it of the required size, cemented to the glass plate, makes an excellent guide for the drill. The tube can be rotated by the fingers, or if a wooden pulley is attached to it, the common drill-stock bow may be used. If the drill is moistened with a paste of fluor-spar and sulphuric acid, the labor of drilling glass is much lessened.

606. Drawing on Glass. — Drawings in India-ink for lantern projections are easily executed on glass by first flowing the plate with a solution of plain collodion, and allowing it to dry. As the transparency is not in the least affected, the plate may be placed directly over the picture to be reproduced, and traced in India-ink with a fine steel pen, even by one having little or no artistic skill.

607. Useful Cements. — Frequent use will be found for reliable cements. The following are confidently recommended: —

For cementing wood, leather, metal, or glass to glass, melt at 100° C. one part of gutta-percha and one part of pine-pitch, stirring till homogeneous. Soften the cement by heat when needed for use.

For an acid-proof cement, make a concentrated solution of sodium silicate, and form a paste, with powdered glass.

For work in electricity an excellent insulating cement is made by melting together rosin, $3\frac{1}{2}$ lbs., beeswax, $2\frac{1}{2}$ lbs., Venetian-red, 2 lbs., and Venice turpentine, 12 oz. Less Venice turpentine will make it harder.

608. Silvering Glass. — Dissolve 154 grains of silver nitrate in 17 fluid ounces of distilled water. Add ammonia water until the precipitate formed is nearly redissolved. Filter and add distilled water, so as to make the whole 34 fluid ounces. This gives solution A. Thirty-one grains of silver nitrate are dissolved in 34 fluid ounces of boiling distilled water; dissolve 23 grains of Rochelle salt in a small quantity of water, add it to the boiling nitrate, boil till the precipitate becomes gray. Filter and allow to cool. This gives solution B. Clean the glass object perfectly by rubbing it with strong nitric acid, using a stick for the purpose. Wash off the acid, then wash the article with caustic potash, then with alcohol, and finally with distilled water. Place the object, with the side uppermost which has to be silvered, in a clean dish, and while still wet pour over it, and mix equal quantities of the solutions A and B, covering it half an inch. In about two hours, according to the temperature, the silvering is complete; the hotter the quicker; the slower the better. The object is now taken out, cleaned of its superfluous silver, dried, and varnished. If the coating is thick, the silver surface may be polished with rouge powder on cotton-wool.

609. Cleaning Mercury.— Mercury is an absolute necessity in many lines of investigation. During its use it is liable to become alloyed with zinc or tin, or both, rendering it unfit for the service required of it in experimenting. To remove these metals proceed as follows: To every 100 parts of mercury add 10 parts of a strong solution of ferric chloride and 100 parts of water. Shake in a strong bottle till thoroughly mixed, the mercury becoming broken up into small globules. Set aside for two or three days in a cool place, then decant off the liquid from the mercury, and afterwards wash thoroughly by shaking it up again with dilute hydrochloric acid, and finally with hot water. Remove with porous paper as much of the water as possible; then pour into a filter having a small pin-hole in the bottom, and finally dry over a water-bath.

610. Soldering.— Prepare a soldering fluid by adding scraps of zinc to hydrochloric acid till it refuses to dissolve any more. Wet the surfaces to be joined with this fluid, place between them a few pieces of soft solder, an alloy of lead and tin; then apply heat either by holding the article in a flame, by touching it with a heated soldering-iron till the solder melts, or by directing upon it a blow-pipe flame; on cooling, the surfaces will be firmly joined together.

611. Amalgamating Battery-Zincs. — Dissolve 15 ccm. of mercury in a mixture of 170 ccm. of nitric acid and 625 ccm. of hydrochloric acid. Keep the liquid in a glass-stoppered bottle. To use, immerse the zinc in the fluid for a few minutes, then remove, and rinse thoroughly in water.

612. Battery Fluids for the Carbon and Zinc Battery.

No. 1. — Water, 8 kilo.; pulverized potassium bichromate, 1.2 kilo.; sulphuric acid, 3.6 kilo. Put the powdered potassium bichromate in the water, and when dissolved add the acid in a fine stream, with constant agitation.

No. 2. — Dissolve 4 oz. of chromic acid in one quart of dilute sulphuric acid, 1 vol. sulphuric acid to 12 of water, and allow the solution to cool and settle.

The last of these solutions is greatly to be preferred to the former, as the absence of potassium prevents the formation of chrome alum crystals, the defect of the first solution. Chromic acid can now be obtained at greatly reduced prices, and produces a solution lasting much longer than that prepared from the bichromate.

As the chromic acid solution is frequently used as a depolarizer in the Bunsen battery, the following preparation from chromic acid will be found very efficient: Dissolve one pound of chromic acid in one part of water, and add seven fluid ounces of sulphuric acid, with stirring. It is claimed that the constancy is improved by adding one-third volume of nitric acid.

613. Care of Batteries. — All the connections in a battery must be kept clean and bright, and the zincs well amalgamated. The porous cups must be thoroughly soaked in water before using, and kept in water while the battery is idle. In the Daniell's battery they become clogged in time with copper, and hence useless. They can be renovated by soaking in nitric acid, but it is probably just as cheap to throw them away and substitute new ones. In the carbon battery, where the potassium bichromate solution is used, the carbons become clogged with chrome alum crystals and chromium oxide, greatly increasing the resistance. If the carbons are soaked for an hour in nitric acid the difficulty will be removed. Chromium oxide is formed in the case of the chromic acid solution, and in time affects the action of the battery.

It often happens that the carbons become detached from the brass clamp in the case of the Grenet battery, and need to be resoldered, a thing easily done if the end of the carbon is first plated with copper. To do this, connect to the negative pole of a Daniell's battery the carbon plate, and to the other a strip of copper. Let these dip close together, but not in contact, in a vessel containing a solution of copper sulphate, removing them when the plating covers completely the surface to be soldered. Cover with beeswax the surface not to be plated.

614. Electrical Amalgam. — Take two parts by weight of tin, one part of zinc, and eight parts of mercury. Melt the tin in an iron ladle, add the zinc, and raise it to the melting-point. Now add the mercury, and stir the mixture till it is cool. Apply to the rubber by mixing with the amalgam a little lard.

615. Cleaning Electrical Machines. — To remove dust from electrical machines, use a cloth wet with kerosene oil. The efficiency of both the Frictional and the Holtz Machine is frequently largely increased when this is done. Even the old form of Holtz machine seldom fails to work when treated in this way.

APPENDIX C.

TABLES FOR REFERENCE.

Table I.

CAPILLARITY.

Height to which the liquid will rise in a tube 1 mm. in diameter: —

Alcohol	11.4 mm.	Mercury	.—9.2 mm.
Bromine	9.0	Turpentine	12.7
Ether	9.5	Water	29.3

Table II.

DENSITIES OF VARIOUS SUBSTANCES.

Acetic acid	1.060	Benzine	0.72 to 0.740
Agate	2.615	Benzole	0.899
Alcohol, absolute	0.806	Birch	0.690
Alcohol, common	0.833	Bismuth, cast	9.822
Alum	1.724	Blood	1.060
Aluminium	2.670	Boxwood	1.280
Amber	1.078	Brass, cast	8.400
Antimony, cast	6.720	Brass, sheet	8.440
Apple-tree wood	0.790	Brick	1.6 to 2.000
Arsenic	8.310	Bromine	3.187
Ash, dry	0.690	Butter	0.942
Ash, green	0.760	Calcium chloride	2.230
Asphalt	2.500	Camphor	0.988
Basalt	2.950	Carbon disulphide	1.293
Beech, dry	0.690 to 0.800	Carbon dioxide, liquid	0.947
Beeswax	0.964	Cedar, American	0.554
Bell-metal	8.050	Chalk	1.8 to 2.800

Cherry-tree	0.710		Ice	0.917
Chestnut	0.606		Iceland spar	2.723
Chloroform	1.525		Iron, bar	7.788
Clay	1.920		Iron, cast	7.230
Coal, anthracite	1.26 to 1.800		Iron, wrought	7.780
Coal, bituminous	1.270 to 1.423		India-rubber	0.930
Cobalt	8.800		Iodine	4.950
Concrete, ordinary	1.900		Iron pyrites	5.000
Concrete, in cement	2.200		Ivory	1.820
Cork	0.240		Lard	0.947
Copper, cast	8.830		Lead, cast	11.360
Copper, sheet	8.878		Lead, sheet	11.400
Deal, Norway	0.689		Lignum vitae	1.333
Diamond	3.530		Lime, quick	0.843
Earth	1.520 to 2.000		Limestone	3.180
Ebony	1.187		Logwood	0.913
Elder	0.690		Magnesium	1.750
Elm	0.579		Mahogany	0.56 to 0.852
Elm, Canadian	0.725		Maple	0.755
Emery	3.900		Marble	2.720
Ether	0.736		Mercury	13.596
Emerald	2.770		Milk	1.032
Feldspar	2.600		Molasses	1.426
Fir, spruce	0.512		Mortar, average	1.700
Fluor-spar	3.200		Naphtha	0.848
Galena	7.580		Nitric acid	1.38 to 1.559
German-silver	8.432		Oak, American red	0.850
Glass, flint	3.000 to 3.600		Oak, American white	0.779
Glass, crown	2.520		Oak, live, seasoned	1.068
Glass, plate	2.760		Oak, live, green	1.260
Glycerine	1.260		Oil, castor	0.970
Gold	19.360		Oil, linseed	0.940
Gypsum, crys.	2.310		Oil, olive	0.915
Granite	2.650		Oil, turpentine	0.870
Graphite	2.500		Oil, whale	0.923
Gun-metal	8.561		Paraffine	0.824 to 0.940
Gutta-percha	0.966		Petroleum	0.836
Heavy spar	4.430		Phosphorus	1.830
Honey	1.450		Pear-tree	0.660
Human body	0.890		Pine, red, dry	0.590
Hydrochloric acid, aq. sol.	1.222		Pine, white, dry	0.554

Pine, yellow, dry	0.461	Steel, unhammered		7.816
Pine, pitch	0.660	Sugar, cane		1.593
Pitch	1.150	Sulphur, native		2.033
Platinum wire	21.531	Sulphuric acid		1.840
Poplar, common	0.389	Tallow		0.940
Porcelain, china	2.380	Tar		1.015
Potassium	0.865	Tin, cast		7.290
Quartz	2.650	Tourmaline, green		3.150
Rock-salt	2.257	Vinegar		1.026
Saltpetre	2.100	Water, at 100° C.		0.958
Sand, quartz	2.750	Walnut		0.680
Sand, river	1.880	Water, sea		1.027
Sand, fine	1.520	Wax, white		0.970
Sand, coarse	1.510	White metal, Babbitt		7.310
Silver, cast	10.424 to 10.511	Willow		0.585
Slate	2.880	Zinc, cast		7.000
Sodium	0.970			

The above table gives the weight in grammes of 1 ccm. of the substance. It should be considered as giving only approximations, as most of the densities vary between wide limits in different specimens.

GASES AND VAPORS AT 0° C., BAROMETRIC PRESSURE OF 76 CM.

Acetic acid vapor	2.0080	Hydrogen disulphide	1.1912
Air	1.0000	Hydrogen	0.0693
Alcohol vapor	1.6138	Hydriodic acid	4.3737
Ammonia	0.5967	Marsh gas	0.5540
Benzole vapor	2.7290	Mercury vapor	6.9207
Carbonous oxide	0.9670	Nitrogen	0.9714
Carbon dioxide	1.5241	Nitrogen binoxide	1.0392
Chlorine	2.4501	Nitrogen monoxide	1.5241
Coal-gas	0.340 to 0.6500	Oxygen	1.1057
Cyanogen	1.8060	Sulphurous acid	2.2113
Ether vapor	2.5630	Water vapor	0.6225
Hydrochloric acid	1.2780		

In the above table the unit adopted is the quantity of matter in 1 ccm. of air under the standard conditions.

Table III.

LIMIT OF ELASTICITY.

Cast steel, drawn	55.6	Platinum, drawn	26.0
" " annealed	5.0	" annealed	14.5
Copper, drawn	12.0	Steel, drawn	42.5
" annealed	3.0	" annealed	15.0
Iron, drawn	32.5	Silver, drawn	11.25
" annealed	5.0	" annealed	2.75
Lead, drawn	0.25	Tin, drawn	0.45
" annealed	0.2	" annealed	0.20

The above table gives the weight in kilogrammes necessary to cause permanent elongation in wires 1 mm. in diameter at the ordinary temperature.

Table IV.

ELECTRICAL CONDUCTIVITY.

Arsenic	4.76	Iron, annealed	15.47
Antimony, pressed	4.24	Lead, pressed	7.67
Aluminium, annealed	51.64	Mercury, liquid	1.58
Bismuth, pressed	1.15	Nickel, annealed	12.07
Brass	26.22	Platinum, annealed	16.61
Cadmium	22.37	Silver, annealed	100.0
Copper, annealed	94.19	Sodium	72.43
German silver	7.14	Tin, pressed	11.39
Gold, annealed	71.83	Zinc, pressed	26.73

The above table expresses the conductivity of the substances relatively to silver at 0° C.

Table V.

APPROXIMATE ELECTRO-MOTIVE FORCE OF PRIMARY BATTERIES.

Daniell's, amalgamated zinc, $H_2SO_4 + 4Aq.$, $CuSO_4$, con. sol. 1.079 volts.
" " " " $+12$ " " " 0.978 "
" " " " $+12$ Aq., $CuNO_3$, " 1.000 "
" equi-dense solutions of $ZnSO_4$, and $CuSO_4$ plates
of pure Zn and Cu . . . . 1.104 "
Bunsen, amal. zinc, $H_2SO_4 + 12$ Aq., and HNO_3, carbon . . 1 964 "
Grove, " " " $+ 4$ Aq., HNO_3, platinum 1.956 "
Leclanché, zinc in saturated solution of NH_4Cl . . . . 1.32 "
Potassium bichromate . . . . 2.00 "

Table VI.

ELECTRICAL RESISTANCE, DIAMETER, CROSS-SECTION, ETC., OF PURE COPPER WIRE, BIRMINGHAM GAUGE, TEMPERATURE 15° C.

No.	Diameter.		Area of Cross-Sec.		Resistance.		Weight.	
	Ins.	Cms.	Sq. Ins.	Sq. Cms.	Ohms. per Yd.	Ohms. per M.	Lbs. per Yd.	Grms. per M.
0000	.454	1.1530	.1620000	1.0444000	.000152	.000167	1.884000	934.7000
000	.425	1.0790	.1420000	.9150000	.000174	.000190	1.651000	819.1000
00	.380	.9650	.1130000	.7320000	.000217	.000238	1.320000	654.8000
0	.340	.7620	.0908000	.5860000	.000272	.000297	1.056000	524.2000
1	.300	.7210	.0707000	.4560000	.000349	.000382	.822000	408.1000
2	.284	.7210	.0693000	.4090000	.000389	.000425	.737000	365.8000
3	.259	.6580	.0527000	.3400000	.000468	.000512	.613000	304.2000
4	.238	.6050	.0445000	.2870000	.000554	.000606	.518000	256.9000
5	.220	.5590	.0380000	.2450000	.000649	.000709	.442000	219.5000
6	.203	.5160	.0324000	.2090000	.000762	.000833	.377000	186.9000
7	.180	.4570	.0254000	.1640000	.000969	.001060	.296000	146.9000

No.	Diameter.		Area of Cross-Sec.		Resistance.		Weight.	
	Ins.	Cms.	Sq. Ins.	Sq. Cms.	Ohms. per Yd.	Ohms. per M.	Lbs. per Yd.	Grms. per M.
8	.165	.4190	.0214000	.1380000	.001150	.001260	.249000	123.5000
9	.148	.3760	.0172000	.1110000	.001430	.001570	.200000	99.3000
10	.134	.3400	.0141000	.0910000	.001750	.001910	.164000	81.4000
11	.120	.3050	.0113000	.0730000	.002180	.002380	.132000	65.5000
12	.109	.2770	.0093300	.0602000	.002640	.002890	.109000	53.9000
13	.095	.2410	.0070900	.0457000	.003480	.003800	.082500	40.9000
14	.083	.2110	.0054100	.0349000	.004560	.004980	.063000	31.2000
15	.072	.1830	.0040700	.0263000	.006060	.006620	.047400	23.5000
16	.065	.1650	.0033100	.0214000	.007480	.008130	.038600	19.2000
17	.058	.1470	.0026400	.0170000	.009330	.010200	.030700	15.3000
18	.049	.1240	.0018900	.0122000	.013100	.014300	.022000	10.9000
19	.042	.1070	.0013900	.0089400	.017800	.019600	.016100	8.0000
20	.035	.0889	.0009620	.0062100	.025600	.028000	.011200	5.5600
21	.032	.0813	.0008040	.0051900	.030700	.033500	.009360	4.6400
22	.028	.0711	.0006160	.0039700	.040000	.043810	.007160	3.5500
23	.025	.0635	.0004910	.0031700	.050200	.054900	.005710	2.8300
24	.022	.0559	.0003800	.0024500	.064900	.070900	.004420	2.1900
25	.020	.0508	.0003140	.0020300	.078600	.085800	.003670	1.8200
26	.018	.0457	.0002540	.0016400	.096900	.106000	.002960	1.4700
27	.016	.0406	.0002010	.0013000	.123000	.134000	.002340	1.1600
28	.014	.0356	.0001540	.0009930	.160000	.175000	.001790	0.8890
29	.013	.0330	.0001330	.0008560	.186000	.203000	.001540	.7660
30	.012	.0305	.0001130	.0007320	.218000	.238000	.001320	.6530
31	.010	.0254	.0000785	.0005070	.314000	.343000	.000915	.4540
32	.009	.0229	.0000636	.0004100	.388000	.424000	.000746	.3670
33	.008	.0203	.0000503	.0003240	.491000	.536000	.000585	.2900
34	.007	.0178	.0000385	.0002480	.641000	.701000	.000442	.2200
35	.005	.0127	.0000196	.0001270	1.260000	1.370000	.000229	.1130
36	.004	.0102	.0000126	.0000811	1.960000	2.150000	.000146	.0720

TABLES FOR REFERENCE.   367

Table VII.

ACCELERATION DUE TO GRAVITY.

Berlin,   lat. 52°30' . 981.25 cm.   Washington, lat. 38°54' . 980.06 cm.
Greenwich, " 51°29' . 981.17 "   Lat. of 45° . . . . . 980.61 "
Paris,   " 48°50' . 980.94 "   Equator . . . . . . 978.10 "
New York, " 40°43' . 980.19 "   Pole . . . . 983.11 "

Table VIII.

HEAT, ABSORBING, CONDUCTING, RADIATING, REFLECTING POWER.

Substance.	Absorbing.	Conducting.	Radiating.	Reflecting.	Substance.	Absorbing.	Conducting.	Radiating.	Reflecting.
Bismuth		1.8			Lead, white	100		100	
Brass, polished		33.1	7	100	Mercury			20	
Copper		77.6	7		Palladium		6.3		
Glass, white			90	10	Paper			98	
Gold, polished		53.2	3		Platinum		8.4	17	
Indian Ink	85		85	13	Rose's alloy		2.8		
Iron		11.9			Shellac	72		72	
Isinglass	91		80		Silver, polished		100	3	90
Lampblack	100		100	0	Steel		12.0		70
Lead, polished		8.5	19	60	Tin		14.5	12	
" tarnished			45		Zinc		19.9		

In the table of absorbing powers the standard adopted is lampblack and the source of heat is copper at 100° C. In conducting powers, silver is the standard. In radiating power, lampblack is taken as the standard. The initial temperatures of the substances compared is 100° C. In reflecting powers, polished brass is taken as the standard.

Table IX.
BOILING POINTS OF SUBSTANCES AT BAR. PRES. 76 CM.

Acetic acid	119.00° C.	Petroleum	40 to 70.0° C.
Acetone	56.28	Phosphorus	290.0
Alcohol	78.39	Sulphur	448.4
Aldehyde	20.78	Sulphuric acid	338.0
Ammonia	—39.00	Suphur dioxide	—8.0
Amylic alcohol	131.00	Turpentine	159.3
Aniline	183.70	Water, distilled	100.0
Benzine	90 to 110.00	Water, saturated with NaCl	102.0
Benzole	80.44		
Bromine	63.00	Water, saturated with KNO_3	116.0
Carbon dioxide	—78.00		
Carbon disulphide	48.00	Water, saturated with K_2CO_3	135.0
Ether	34.90		
Iodine	200.00	Water, saturated with $CaCl_2$	179.0
Mercury	350.00		
Nitrous oxide	—92.00	Zinc	1040.0
Methylic alcohol	65.5		

Table X.
COEFFICIENTS OF EXPANSION FOR 1° BETWEEN 0° AND 100° C.
LINEAR.

Aluminium	0.00002221	Lead	0.00002799
Antimony	0.00000980	Marble	0.00000786
Bismuth	0.00001330	Paraffine	0.00027854
Brass	0.00001875	Pine	0.00000496
Bronze	0.00001844	Platinum	0.00000886
Copper	0.00001866	Sandstone, red	0.00001174
Ebonite	0.00008420	Silver	0.00001943
Glass	0.00000861	Sulphur	0.00006413
Gold	0.00001466	Steel, tempered	0.00001322
Graphite	0.00000786	" untempered	0.00001095
Iron, cast	0.00001125	Tin	0.00002730
" wrought	0.00001220	Zinc	0.00002976

TABLES FOR REFERENCE. 369

CUBICAL.

Air, cons. vol., 0°–100° . 0.003663	Ether, 0°–63° 0.002100
" cons. pres., 0°–100° 0.003667	Mercury, 0°–100° . . . 0.000181
Alcohol, ethyl. 0°–50° . 0.001080	Sulphuric acid, 0°–30° . 0.000489
" methyl. 0°–61°. 0.001358	Turpentine, —9°–106° . 0.001050
Benzine, 11°–81° . . . 0.001385	Water, 4°–100° 0.000449

Table XI.

LATENT HEAT OF LIQUEFACTION AND VAPORIZATION.

LIQUEFACTION.

Beeswax 97.22	Silver 21.07
Bismuth 12.64	Sodium nitrate 62.97
Lead 5.37	Sulphur 9.37
Mercury 2.83	Tin 14.25
Phosphorus 5.03	Water 79.24
Platinum 27.18	Zinc 28.13
Potassium nitrate . . . 47.37	

VAPORIZATION.

Alcohol, ethylic 202.4	Ether 90.4
" methylic 264	Mercury 62.0
Acetic acid 102	Oil of turpentine 74.0
Bromine 45.6	Water 535.9
Carbon disulphide . . . 86.7	

Table XII.

MELTING POINTS

Antimony . . . . . 425° C.	Iridium . . . . . 1950.0° C.
Beeswax . . . . . 62°	Iron . . 1500.0° to 1600°
Bismuth . . . . . 270°	Lard . . . . 33.2°
Brass . . . . . . 1020°	Lead . . . . 334°
Bromine . . . . . —24.5°	Margaric acid . . . 59.9°
Butter . . . . . . 33.0°	Mercury . . . . . —38.8°
Copper . . 1054.0° to 1200°	Paraffine . . . . 38°–52°
German-silver . . . 1093°	Phosphorus . . . . 44.2°
Gold . . . . 1250.0°	Platinum . . 1775° to 2000°
Ice . . . . 0.0°	Potassium . . . . 62.5°

370 PRACTICAL PHYSICS.

Rose's metal	94° C.	Tallow (fresh)	43° C.
Silver	1000°	Tin	235°
Sulphur	115°	Turpentine	—27°
Sodium	97.6°	Wax, white	65°
Stearic acid	69.9°	Wood's metal	68°
Stearine	60°	Zinc	433°
Spermaceti	49°		

Table XIII.

SPECIFIC HEAT.

Alcohol at 0°	0.5475	Mercury	0.0335
Aluminium	0.2122	Nitrogen	0.2438
Antimony	0.0486	Olive oil	0.310
Bismuth	0.0298	Oxygen	0.2175
Brass, hard	0.0858	Platinum	0.0323
Carbon disulphide	0.2352	Silver	0.0559
Copper	0.0933	Salt	0.173
Ether at 0°	0.5290	Sulphur	0.1844
Glass, thermometer	0.1980	Steel	0.118
Glycerine	0.5550	Tin	0.0559
Ice	0.5040	Turpentine at 0°	0.4106
Iron	0.1124	Wood spirit	0.645
Lead	0.0315	Zinc	0.0935

Table XIV.

INDICES OF REFRACTION.

Air	1.000294	Glycerine	1.47
Aqueous lens, eye	1.357	Ice	1.310
Alcohol, ethyl	1.361	Iceland spar, ordinary ray	1.654
Benzine	1.49	" " extra. ray	1.483
Canada balsam	1.54	Quartz, ordinary ray	1.544
Carbon disulphide	1.626	" extra. ray	1.553
Crown glass	1.515	Rock salt	1.550
Crystalline lens, eye	1.384	Ruby	1.779
Diamond	2.47 to 2.75	Vitreous lens, eye	1.339
Ether	1.354	Water	1.332
Flint glass	1.57 to 1.71		

These indices are given for the mean D line.

TABLES FOR REFERENCE. 371

Table XV.

MENSURATION RULES.

Area of triangle = base × ½ altitude.
" parallelogram = base × altitude.
" trapezoid = altitude × ½ sum of parallel sides.
Circum. of circle = diameter × 3.1416.
Diameter of circle = circum. × .3183.
Area of circle = diameter squared × .7854.
Area of ellipse = product of diameters × .7854.
Area of reg. polygon = sum of sides × ½ apothem.
Lat. surface of cylinder = cir. base × alt.
Contents of cylinder = area base × alt.
Surface of sphere = diameter × circum.
Contents of sphere = diameter cubed × .5236.
Surface of pyramid ⎫
" " cone ⎬ = cir. base × ½ slant height.
Contents of cone = area base × ⅓ alt.
Surface of frustum of pyramid or cone =
    sum of cir. of bases × ½ slant height.
Contents of frustum of pyramid or cone =
⅓ alt. × sum of areas of bases and sq. rt. of product of these areas.

Table XVI.

LENGTH OF SECONDS' PENDULUM.

Greenwich, lat. 51° 29' . . . . 99.413 cm.
Paris,   48° 50' . . . . 99.390 "
New York,   40° 43' . . . . 99.317 "
Washington,   38° 54' . . . . 99.306 "
    Lat.   45° 00' . . . . 99.356 "
    Equator . . . . 99.103 "
    Pole . . . . 99.610 "

Table XVII.

VELOCITY OF SOUND AT 0° C.

Air	per sec.,	1,093 ft.	Hydrogen	per sec.,	4,163 ft
Ash	"	15,314	Iron	"	17,822
Brass	"	10,885	Lead	"	4,030
Caoutchouc	"	197	Maple	"	13,472
Carbon monoxide	"	1,106	Oak	"	12,622
Carbon dioxide	"	856	Oxygen	"	1,040
Cedar	"	16,503	Pine	"	10,900
Chlorine	"	677	Silver	"	8,553
Copper	"	11,666	Steel	"	17,182
Elm	"	13,516	Tallow	"	1,170
Ether	"	3,801	Turpentine at 24°	"	3,976
Fir	"	15,218	Walnut	"	15,095
Glass	"	16,488	Water at 8.1°	"	4,708
Gold	"	5,717	Wax	"	2,811

Table XVIII.

ELASTICITY OF TRACTION.

No. of kilos. required to double the length of a wire 1 sq. mm. in section.

Brass	9,000	Silver	7,400
Copper	12,400	Slate	11,035
Glass, plate	7,015	Steel	21,000
Iron, wrought	19,000	Whalebone	700
Lead	1,800	Wood	1,100
Platinum	17,044	Zinc	8,700

Table XIX.
TENACITY.

Copper, drawn	40.3	Silver, drawn	29.00
" annealed	30.54	" annealed	16.02
Iron, drawn	61.10	Steel, drawn	70.00
" annealed	46.88	" annealed	40.00
Lead, drawn	2.07	" cast, drawn	80.00
" annealed	1.80	" " annealed	65.75
Platinum, drawn	34.10	Tin, drawn	2.45
" annealed	23.5	" annealed	1.70

The above table gives the weight in kilogrammes required to break a wire of the substance 1 mm. in diameter.

Table XX.
TRIGONOMETRICAL FUNCTIONS.

Angle.	Sine.		Tangent.		Cotangent.		Cosine.		Angle.
0	0.000		0.000				1.000		90
		17		17				0	
1	0.017		0.017		57.29		1.000		89
		18		18				1	
2	0.035		0.035		28.64		0.999		88
		17		17				0	
3	0.052		0.052		19.08		0.999		87
		18		18				1	
4	0.070		0.070		14.30		0.998		86
		17		17				2	
5	0.087		0.087		11.43		0.996		85
		18		18				1	
6	0.105		0.105		9.514		0.995		84
		17		18				2	
7	0.122		0.123		8.144		0.993		83
		17		18				3	
8	0.139		0.141		7.115		0.990		82
		17		17		841		2	
9	0.156		0.158		6.314		0.988		81
		18		18		643		3	
10	0.174		0.176		5.671		0.985		80
		17		18		526		3	
11	0.191		0.194		5.145		0.982		79
		17		19		440		4	
12	0.208		0.213		4.705		0.978		78
		17		18		374		4	
13	0.225		0.231		4.331		0.974		77
		17		18		320		4	
14	0.242		0.249		4.011		0.970		76
		17		19		279		4	
15	0.259		0.268		3.732		0.966		75
		17		19		245		5	
Angle.	Cosine.		Cotangent		Tangent.		Sine.		Angle.

Angle.	Sine.		Tangent.		Cotangent.		Cosine.		Angle.
16	0.276		0.287		3.487		0.961		74
		16		19		216		5	
17	0.292		0.306		3.271		0.956		73
		17		19		193		5	
18	0.309		0.325		3.078		0.951		72
		17		19		174		5	
19	0.326		0.344		2.904		0.946		71
		16		20		157		6	
20	0.342		0.364		2.747		0.940		70
		16		20		142		6	
21	0.358		0.384		2.605		0.934		69
		17		20		130		7	
22	0.375		0.404		2.475		0.927		68
		16		20		119		6	
23	0.391		0.424		2.356		0.921		67
		16		21		110		7	
24	0.407		0.445		2.246		0.914		66
		16		21		101		8	
25	0.423		0.466		2.145		0.906		65
		15		22		95		7	
26	0.438		0.488		2.050		0.899		64
		16		22		87		8	
27	0.454		0.510		1.963		0.891		63
		15		22		82		8	
28	0.469		0.532		1.881		0.883		62
		16		22		77		8	
29	0.485		0.554		1.804		0.875		61
		15		23		72		9	
30	0.500		0.577		1.732		0.866		60
		15		24		68		9	
31	0.515		0.601		1.664		0.857		59
		15		24		64		9	
32	0.530		0.625		1.600		0.848		58
		15		24		60		9	
33	0.545		0.649		1.540		0.839		57
		14		26		57		10	
34	0.559		0.675		1.483		0.829		56
		15		25		55		10	
35	0.574		0.700		1.428		0.819		55
		14		27		52		10	
36	0.588		0.727		1.376		0.809		54
		14		27		49		10	
37	0.602		0.754		1.327		0.799		53
		14		27		47		11	
38	0.616		0.781		1.280		0.788		52
		13		29		45		11	
39	0.629		0.810		1.235		0.777		51
		14		29		43		11	
40	0.643		0.839		1.192		0.766		50
		13		30		42		11	
41	0.656		0.869		1.150		0.755		49
		13		31		39		12	
42	0.669		0.900		1.111		0.743		48
		13		33		39		12	
43	0.682		0.933		1.072		0.731		47
		13		33		36		12	
44	0.695		0.966		1.036		0.719		46
		12		34		36		12	
45	0.707		1.000		1.000		0.707		45
Angle.	Cosine.		Cotangent		Tangent.		Sine.		Angle.

TABLES FOR REFERENCE. 375

Table XXI.

SOME USEFUL NUMBERS.

$\pi = 3.1415926$.
Dyne in grammes $= .00102$.
Poundal in dynes $= 13825$.
Erg in gramme-centimetres $= .00102$.
Foot-pound in kilogramme-metres $= .13825$.
Kilogramme-metre in foot-pounds $= 7.23308$.
Foot-poundal in ergs $= 421390$.
$\sqrt{2} = 1.4142$.
$\sqrt{3} = 1.7320$.
$\sqrt{5} = 2.2361$.
A cubic foot of water at 4° C.
  weighs in pounds $= 62.425$.
A cubic foot of water at $16\frac{2}{3}^{\circ}$ C.
  weighs in pounds $= 62.321$.

A cubic foot of air at 0° C.
  weighs in pounds $= 0.080728$.
1 litre of H. at 0° C., 760 mm.,
  weighs 0.08969 g.
1 Paris foot $= 0.32484$ metres.
1 " line $= 2.2588$ mm.
1 Eng. foot $= 0.30479$ m.
1 Ger. mile $= 7.4204$ kilom.
1 Eng. mile $= 1.60929$ kilom.
1 Rhenish ft. $= 0.31385$ m.
1 metre $= 3.2809$ Eng. ft.
1 kilometer $= 0.62138$ Eng. mile.
1 litre $= 0.22017$ gal.
1 " $= 1.76133$ pints.
1 kilogramme $= 2.20462$ lbs.avoir.
1 gramme $= 15.43235$ grains.
1 metre $= 39.37$ in.
1 U. S. gal. $= 231$ cu. in.

Table XXII.

WEIGHTS AND MEASURES.

MEASURES OF LENGTH, ENGLISH.

1 mi. $= 8$ fur. $= 320$ rods $= 1760$ yd. $= 5280$ ft. $= 63360$ in.

MEASURES OF LENGTH, FRENCH.

1 kilo. $= 1000$ m. $= 10000$ dcm. $= 100000$ cm. $= 1000000$ mm.

MEASURES OF SURFACE, ENGLISH.

1 acre $= 4840$ sq. yd. $= 43560$ sq. ft.

MEASURES OF SURFACE, FRENCH.

1 sq. H km. $= 10$ sq. D km. $= 100$ sq. m. $= 1000$ sq. dcm. $= 10000$ sq. cm.
  $= 100000$ sq. mm.

1 are $= 100$ sq. metres.

Measures of Volume, English.

1 cu. yd. = 27 cu. ft. = 46656 cu. in.

Measures of Volume, French.

1 cu. metre = 1000 cu. dcm. = 1000 litres = 1000000 ccm.

English Weights.

1 lb. avoir. = 16 oz. = 256 dr. = 7000 gr.

1 oz. = 437.5 gr.

French Weights.

1 kilo. g. = 1000 g. = 10000 dcg. = 100000 cg. = 1000000 mg.

Miscellaneous.

Lineal feet × .00019 = miles.
Square inches × .007 = sq. ft.
Cu. inches × .00058 = cu. ft.
Cu. ft. × 7.48 = U. S. gallons.
Cu. in. × .004329 = U. S. gallons.
U. S. gals. × .13367 = cu. ft.
Cu. ft. of water × 62.5 = lbs. avoir.
Cu. in. of water × .03617 = lbs. avoir.
Metres × 3.2809 = ft.
Ft. × 0.3048 = metres.
Sq. in. × 6.451 = scm.
Scm. × 0.155 = sq. in.
Cu. in. × 16.386 = ccm.
Ccm. × .06103 = cu. in.
Litres × 61.027 = cu. in.
Oz. avoir. × 28.35 = grammes.
Lb. × 453.593 = grammes.
Gr. × 15.432 = grains.
Kilog. × 2.2046 = lbs. avoir.

INDEX.

INDEX.

[The numbers refer to the articles.]

Aberration, spherical, 537, 548.
Absorption, of air by water, 52; of heat by substances, table, page 367.
Acceleration due to gravity, 131; table, page 367.
Adhesion, 64, 65.
Air-pump, experiments with, 152-154, 160-162, 178, 185, 251, 261; mercury, 163.
Air-thermometer, 240, 258.
Amalgam, electric, 614.
Amalgamating battery-zincs, 611.
Analysis of white light, 551.
Aneroid barometer, 159.
Angle, critical, 541; measurement of, 18; of prism, 522.
Angular currents, laws of, 382, 383.
Apparatus for laboratory, list of, 595.
Archimedes, principle of, 180-185.
Area, measurement of, 18, 19.
Ascent of liquids in capillary tubes, 85; between plates, 84.
Astatic galvanometer, 358.
Astronomical telescope, 585.
Athermancy, 264.
Atmospheric pressure, 151-162.
Attraction, electrical law of, 312; magnetic law of, 287; of vibrating bodies, 496-498.
Atwood's machine, 130.
Aurora-tube, 402.

Balance, beam, how to use, 23; Jolly's, how to make and use, 23.
Barometer, 158-159.
Baroscope, 185.
Batteries, care of, 613; connecting together, 369; galvanic, 346-348; for laboratory, 595.
Battery, Daniell's, 358, 595; floating, 383; fluid for carbon, 612; Gre1et, 595; measure electromotive force of, 371-373; measure resistance of, 367-369; secondary or storage, 408-409; table of E. M. F. of, page 365; thermo-electric, 406.
Beam-compass, 15.
Beats, 459, 460, 462.
Boiling-point of liquids, 254; influence of substance in solution, 255; of pressure, 251, 252, 254; on thermometers, 246; table of, page 368.
Books of reference for laboratory, 594.
Boyle, law of, 165.
Branch currents, 375.
Bridge, Wheatstone's, 358.
Bunsen's photometer, 516.
Buoyancy of fluids, 180-185.

Calibration of galvanometer, 374, 376.

380   INDEX.

Caliper, inside, 13; micrometer, 9; outside, 12; verniered steel, 6.
Calorimetry, 269–276.
Camphor, motion of, 82.
Candle-power, measurement of, 517, 518.
Capillarity, table of, page 361.
Capillary action, 77–85; laws of, 85; measure diameter of tube, 34.
Carbonic acid, testing room for, 231; velocity of sound in, 435.
Cartesian diver, 184.
Cements, 607.
Centre of mass, 117–119; of oscillation, 139; of percussion, 141.
Charge of Leyden jar, not in metal, 340; residual, 335.
Charging Leyden jars, 335–340.
Chemical effects of electrical currents, 353–355.
Chladni's plates, 491.
Chromic acid solution for batteries, 612.
Circuit, divided, law of, 375.
Cleavage, 70.
Coefficient of expansion, 234, 235, 241; table of, page 368.
Coercive force, 298.
Cohesion, 61–71; figures, 80; of liquids, 64; of solids, 61, 62.
Coil, induction, 394; resistance, 358.
Cold, artificial production of, 277–283.
Collision of bodies, 113–116.
Color, 557–566; complementary, 561–566; mixed, 561, 562; produced by diffraction, 578, 579; by pressing two plates together, 516; by polarized light, 590–593; simple, 551–552; unequally refrangible, 553.
Commutator, to make, 347.
Compound microscope, 584; polarizing attachment, 592.
Concave lenses, 546–548; mirrors, 532, 534, 535, 537.
Condensers, 332–340.
Conduction of heat, 215–223, page 367; electricity, 318, page 364; of sound, 428–432.
Conductometer, 217, 219.
Conductors, distribution of electrification on, 325–331.
Cone, volume of frustum of, 22.
Controlling magnet for galvanometer, 365.
Convection of heat, 225–231.
Convex lenses, 545, 547–549; mirrors, 533, 534, 536.
Cooling, affected by character of surface, 265; Newton's law of, 267.
Corks, boring holes in, 595.
Critical angle, 541.
Crova's disk, 439.
Crystallization, 66–70; increase of volume due to, 242–244.
Cube, Leslie's, 258.
Current electricity, 345–356; extra, 395; induced, 386–397.
Curved mirrors, 532–537.
Curvilinear motion, 125–127.
Cylindrical jar, to measure, 13.

Daniell's battery, 358, 595.
Dark lines of solar spectrum, 568, 569.
Density, determination of, 187–198; of ice, 244; table of, page 361.
Deviation, angle of, 540.

INDEX.

Diagonal scale, 2.
Dialyzer, 101.
Diameter of wires, table of, page 365.
Diapason, 459.
Diathermancy, 264.
Differential thermometer, 258.
Diffraction of light, 578-581.
Diffusion, 95-106; fountain, 102.
Dip of needle, 306.
Direction of current, 347, 348.
Discharger, jointed, 335.
Discord, 487-489.
Dispersion of light, 551.
Distillation, 256.
Distribution of electrification, 324-331; of magnetism, 301.
Divided currents, 375.
Dividers, 2; proportional, 16.
Divisibility of matter, 44-47.
Double refraction, 589.
Drop-size of liquids, 63.
Duration of electric spark, 344.
Dust figures, 425.

Elasticity, 73-76; table of limits of, page 364; of traction, table, page 372.
Electric amalgam, 614; attraction and repulsion, 309-314; conductivity, 318; motor, 388.
Electrical condensers, 333-340; conductivity, 318, page 364; conductivity affected by heat, 364; urrents, direction of, 347, 348; currents, division of, 376; currents, effects of, 350-356; distribution, 325-331; machines, cleaning, 615; measurements, 358-376; pistol, 352; resistance, table of, page 365.

Electricity, 308-409; current, 345-356; developed by chemical action, 346-348; developed by friction, 308-342; devélopéd by heat, 406; developed by induction, 343, 344, 386-397.
Electrification detected, 315-317.
Electrolysis, 353-355.
Electro dynamics, 378-384; magnet, 382; magnetism, 378-384.
Electro-motive force, 370-373; table of, page 365
Electrophorus, 343.
Electroscope, 315.
Equilibrium of forces, 109-112; of liquids, 169; of bodies, 120-122.
Erdmann's float, 20.
Expansion, coefficient of, 234, 239, 241; cubical, 235, 239, 241; table of, page 368; on crystallization, 242-244; of gases, 240, 241; of liquids, 238, 239; of gases producing cold, 207, 281; by heat, 233-244.
Extra current, 395.
Evaporation, cold due to, 279, 280.
Eye, model of, 583.

Falling bodies, law of, 133; independent of mass, 133.
Films, soap, 81, 577.
Fire-syringe, 201.
Flask, specific gravity, 190.
Flexure, elasticity of, 75.
Float, Erdmann's, 20.
Floating battery, 383.
Flotation, principle of, 183.
Fluids, buoyant force of, 181-185; mechanics of, 149-198; pressure in, 150-163.

Focal distance of lenses, 545–546; of spherical mirrors, 535–536.
Force, acceleration due to constant, 129–131; buoyant, 180–185; coercive, 298; electromotive, 371–373; lines of magnetic, 300, 379, 381; measure effect of a, 132.
Forces, composition of, 109–112; parallel, 112.
Fountain, Hero's, 177; intermittent, 176; in vacuo, 162.
Fraunhofer's lines, 568, 569.
Freezing mixtures, 277, 278, 283.
Friction, 142.
Frictional plate machine, 342.
Fusion, latent heat of, 275.

Galilean telescope, 586.
Galvanometer, calibration of, 374, 376; resistance of, 365; simple, to make, 347; tangent, to make, 358.
Gases, cold produced by expansion of, 207, 281; compressibility of, 165; density of, page 363; endosmose of, 102–106; expansibility of, 160; thermal conductivity of, 261; velocity of sound in, 434–436.
Gassiot's cascade, 403.
Glass, bending tube, 602; cutting, 600; closing end of tube, 603; drawing on, 606; drawing out tube, 604; drilling holes in, 605; silvering, 608; smoothing end of tube, 601.
Graphic method, 598; study of sound, 500, 501.
Grating, Nobert's, 578–579.

Gravitation, 129–131; acceleration due to, 131; table, page 367.

Haldat's apparatus, 168.
Hardness, Mohr's scale, 71.
Harmony, 487–489.
Heat, 199–283; absorption of, 263, 264; capacity of substance for, 271–274; capacity of water for, 269, 270; conduction of, 215–223; convection of, 224–231; converted into mechanical motion, 207–210; due to chemical action, 212–214; due to electric currents, 350–352; due to mechanical motion, 207–210; expansion by, 233–244; law of reflection of, 262; latent, of water, 275; latent, of steam, 276; radiant, 258–267; radiation of, affected by surface, 265; radiation of, affected by temperature of surrounding air, 267; specific, 272–274; table of absorbing, conducting, radiating, reflecting power, page 367; table of latent, page 369; table of specific, page 370.
Hemispheres, Magdeburg, 154.
Hero's fountain, 177.
Holtz machine, 344.
Hydrometer, Beaumé's, 196; Nicholson's, 193, 195.

Images, after, 563–566; formed by lenses, 548; formed by mirrors, 528–531, 534; multiple, 530, 531; through small apertures, 511, 512.
Impenetrability, 36–42.
Inclined plane, 148

INDEX. 383

Indestructibility, 58, 59.
Index of refraction, 539, 540; table of, page 370.
Induction, coil, 391, 394; current, 386-397; current on itself, 395; earth, 397; magnetic, 291, 292, 307; statical, 320-323.
Inelastic bodies, 113.
Interference of light, by diffraction, 578-579; by reflection, 576, 577, 581; of sound, 457-462, 492-494.
Intermittent fountain, 176.
Irregular reflection, 523, 524.

Jar, Leyden, 335-340.
Jet, height of, 170, 175, 177.
Jet-tube, to make, 604.
Jolly's balance, 23.
Jurin's laws of capillarity, 85.

König's manometric flames, 502.
Kundt's method of measuring velocity of sound, 437.

Laboratory note-book, 597; operations, 600-615; physical, 594-599; room, 594; rules, 599; work, conducting, 596.
Lantern, magic, 509.
Latent heat, of steam, 276; of water, 275; of liquefaction and vaporization, table, page 369.
Lateral jets, 150.
Lenses, foci of concave, 546; foci of convex, 545; effect on pencils of light, 547; magnifying power of, 549; spherical aberration, 548.

Leslie's cube, 258.
Lever, 144-145.
Leyden jar, charging, 335, 337-339; charging by induction coil, 396; connecting two or more together, 336; discharging, 335; to make, 336; office of coatings, 340; residual charge, 335; spangled, 399.
Liebig's condenser, 256.
Light, 598-593; amount of light reflected, 525, 526; diffused reflection of, 523, 524; dispersion of, 550-556; double refraction of, 538-543; interference of, 575-581; law of intensity of, 515, 516; multiple reflection of, 530, 531; polarization of, 588-593; rectilinear propagation of, 510-513; reflected, amount of, 525, 526; reflection of, 519-526; refraction of, 538-543; regular reflection of, 520-522; single refraction of, 538-543; total reflection of, 541-543.
Linear expansion, coefficient of, 234; table, page 368.
Liquids, buoyancy of, 180-184; conductivity of heat by, 222, 223; diffusion of, 95-101; equilibrium of, 169; expansion of, 228, 229; manner of heating, 225, 226; pressure independent of shape of vessel, 168; specific heat of, 274; transmit pressure, 167.
Lissajou's curves, 503.
Loudness of sound, 447-456.
Luminous effects of electric discharge, 344, 393, 398-404.

384   INDEX.

Machines, electrical, 341–344; simple, 144–148.
Magic lantern, 509; slides, 606.
Magnetic effects of electric currents, 356; field, 300–304; field around conductors carrying current, 378, 379, 381; induction, 291, 292, 307; transparency, 289, 290.
Magnetism, 284–307; nature of, 294–296; terrestrial, 306, 307.
Magnetizing, by currents, 356, 382; by divided touch, 301; by earth, 307; by induction, 292.
Magnetoscope, Coulomb's, 301.
Magnets, effects of heating, 296; effects of jarring, 295; electro, 382; induction by, 291, 292; polarity of, 286–288; to make, 285, 301.
Magnifying-power, 549, 587.
Manometric flames, 502.
Mass, centre of, 118, 119; estimation of, 23–34.
Matter, properties of, 1–106.
Measurement, angular, 17, 18; linear, 2–17; of candle-power, 517, 518; of electro-motive force, 370–373; of effect of force, 132; of focal distance, 535, 536, 545, 546; of magnifying-power, 549, 587; of mass, 23–34; of resistance, 357–369; of surface, 18, 19; of velocity of sound, 434–437; of vibration-number, 472, 473; of volume, 20–22.
Mechanics, of fluids, 149–198; of solids, 107–148.
Melde's experiments with vibrating strings, 477.

Melting-point of substances, 253; table of, page 369.
Mensuration rules, table of, page 371.
Mercury, cleaning, 609.
Micrometer caliper, 9.
Microphone, 390.
Microscope, 584.
Mirrors, curved, 532–537; effect on pencils of light, 532, 533; plane, 528–531; measure focal distance of concave, 535; measure focal distance of convex, 536.
Mixtures, freezing, 283.
Motion, laws of, 108–116; accelerated, 129–131; curvilinear, 124–127; reflection of, 116; wave, 411–417.
Multiple images, 530, 531.

Needle, dipping, 306.
Newton's color disk, 344, 561; law of cooling, 267; rings, 576.
Nicholson's hydrometer, 193, 195.
Nicol prism, 591.
Nobert's grating, 579.
Nodes of organ pipe, 485, 486; of vibrating plates, 491; of vibrating strings, 481.
Norremberg's doubler, to make, 591.
Note-book, laboratory, 597.

Oersted's parallelogram, 347.
Ohm's law, 358.
Optical instruments, 582–587; study of sound, 502, 503.
Optics, 508–593.
Osmose, 99–105.
Overtones, 478–481.

INDEX.

Paradox, 122, 168, 252.
Parallel forces, 112.
Parallelogram of forces, 109–111.
Pascal, law of, 167–170.
Pendulum, Blackburn's sand, 504; law of, 137–141; table of lengths of seconds, page 371.
Percussion, heat developed by, 200.
Photometry, 514–518.
Pitch of sound, 471–473.
Plates, ascension of liquids between, 84; Chladni's, 491; tourmaline, 590; vibration of plates, 491–494.
Point, boiling, 254; affected by pressure, 251–254; affected by substances in solution, 255; located on thermometer, 246.
Polariscope, to make, 591; microscope attachment, 592.
Polarity of magnets, 285-288.
Poles, strength compared, 302.
Polygon, measure area of, 19.
Porosity, 49–54.
Porte lumière, to make, 509.
Powders, density of, 192.
Pressure, depends on depth, 150, 155; effect on boiling-point, 254; elasticity manifested by, 73; exerted by atmosphere, 151–163; exerted in every direction, 154, 155, 167; independent of form of vessel, 168; on a body in a fluid, 180–185.
Principle of Archimedes, 179–185.
Prism, measure angle of, 522; make carbon disulphide, 551; measure refraction of, 540; Nicol, 591.
Projectiles, 134, 135.

Proof-plane, 316.
Propagation of sound, 439.
Protractor, 17.
Pulley, 146; mass determined, 130.
Pump, air, 152–154, 160–162, 178, 185, 251, 261; mercury, 163; suction, 178.
Pyrometer, 209, 234.

Radiant heat, 257–267; causes which modify, 263–266; intensity of, 259; law of reflection, 262.
Radiating power, causes which modify, 265-267.
Radiometer, 210.
Rainbow, project on screen, 556.
Reflection, laws of, 116, 262, 521; irregular, 523–524; regular, 519–522; of heat, 262; of light, 519–526; of sound, 440–443.
Refraction, double, 589; index of, 539, 540; single, 538–543; of sound, 440–446; table of indices, page 370.
Residual charge, 335.
Resistance, coils, 358; of batteries, 367–369; of conductors, 358–364; of galvanometers, 365.
Resonating air-columns, effect of diameter on length of, 454.
Resonators, 455, 456.
Resultant of forces, 109–112.
Rheostat, 358.
Rider, 23.
Rings, Newton's, 576.
Rods, law of vibrating, 475; measure length of, 6.
Rumford's photometer, 517.

Scale, accuracy tested, 15; copy a, 14; diagonal, 2; of hardness, 71.
Sciopticon, 509.
Secondary batteries, 408, 409.
Shadows, 513.
Shunt, 374-376.
Singing flame, 426.
Siphon, 172-177.
Siren, 472.
Soap-film, colors of, 577; solution, to make, 80; strength of, 81.
Soldering, 610.
Solenoid, action of currents on, 382; of magnets on, 382.
Solubility, 87-93.
Sonometer, 476.
Sound, 410-507; attending magnetization and demagnetization, 356; graphic and optical study of, 499-504; interference of, 457-462, 492, 493; loudness of, 447-456; pitch, 471-473; propagation of, 438, 439; reflection of, 440-443; refraction of, 444-446; sources of, 418-426; transmission of, 427-432; velocity of, 433-437; table of velocities, page 372.
Sounding air-columns, 482-486.
Spar, Iceland, 589.
Specific heat, 272-274; table of, page 370.
Spectroscope, 522; to make, 569.
Spectrum, absorption, 571-573; analysis, 567-574; bright line, 570, 574; colors, pure, 552; color unequally refrangible, 553; dark line, 568, 571-573; diffraction, 578-580; mapping, 569; solar, to project on screen, 551.

Sphere, measure diameter of, 12.
Spherical aberration, 537, 548.
Spherometer, 11.
Sprengel's air-pump, 163.
Springs, intermittent, 176.
Stability, 120-123.
Stable equilibrium, 120, 121.
Storage batteries, 408, 409.
Strength of substances, 62.
Strings, laws of, 476, 477.
Suction pump, 178.
Surface tension, 78-82.
Sympathetic vibrations, 463-470.
Syringe, fire, 201.

Table, laboratory, 594; to measure, 5; reference, page 361-376.
Tangent galvanometer, 358.
Telegraph line, 384.
Telephone, acoustic, 430; electric, 389.
Telescope, astronomical, 585; Galilean, 586; magnifying power of, 587; terrestrial, 585.
Temperature of boiling-point of water, 247, 251; of substances, 216, 249.
Tenacity, 62, 74; table of, page 373.
Terrestrial, gravitation, 129-132; magnetism, 305-307; telescope, 585.
Thermal capacity of substances, 272-274.
Thermo-electricity, 406.
Thermometer, accuracy of location of standard points, 246; air-thermometer, 258; comparison with standard, 250; displacement of zero, 248.
Thermometry, 246-256.
Thermopile, 259.

INDEX. 387

Thermoscope, 240.
Thickness, to measure, 9–11.
Tools for laboratory, 594.
Torsion, elasticity of, 76.
Tourmaline tongs, 590.
Traction, elasticity of, table, page 372.
Transmission of sound, 427–432.
Triangle, measure, area of, 18.
Trigonometrical functions, table of, page 373.
Tyndall's experiment on specific heat, 273.

Unstable equilibrium, 121, 122.
Useful numbers, table of, page 375.

Velocity of sound, 433–437; table, page 372.
Ventilation, 227–231.
Vernier, 6.
Vibrating, bodies, attraction of, 496-498; plates and bells, 490–495; rods, 475; strings, 476, 477.
Vibration, number, to measure, 472, 473.
Vibrations, sympathetic, 463–470.

Vocal organs, 505–507.
Volume, of irregular objects, 20; regular objects, 13, 22; substances soluble in water, 21.

Water-hammer, Tyndall's adhesion, 65.
Wave motion, 411–417.
Weighing, manner of, 24; problems in, 25–34.
Weights, centigramme, to make, 33.
Weights and measures, table of, page 375.
Wheatstone's bridge, 358.
Wheel and axle, 147.
Whirling-machine, 125.
White light, decomposition of, 551.
Wire-gauge, 9; measure diameter of, 9; measure resistance of, 357–364.
Wood, thermal conductivity of, 221; transmission of sound by, 429; velocity of sound measured, 437.

Zero, displacement of, 248.
Zinc, amalgamation of, 611.

www.ingramcontent.com/pod-product-compliance
Lightning Source LLC
Chambersburg PA
CBHW022119290426

44112CB00008B/738